Atlas of Developmental Embryology

Also by Emil S. Szebenyi: Atlas of Macaca Mulatta

Atlas

of

Developmental

Embryology

Emil S. Szebenyi

Rutherford - Madison - Teaneck
Fairleigh Dickinson University Press
London: Associated University Presses

© 1977 by Associated University Presses, Inc.
Library of Congress Catalogue Card Number: 75-388

Associated University Presses, Inc.
Cranbury, New Jersey 08512

Associated University Presses
Magdalen House
136-148 Tooley Street
London SE1, 2TT, England

ISBN 0-8386-1710-7
Printed in the United States of America

Contents

Preface

The laboratory study of embryology divides into two major parts, namely, descriptive embryology and experimental embryology. Without study of descriptive embryology, experimental embryology cannot be fully understood. The second one follows the first, as surgery can only follow anatomy. How can someone perform a transplantation experiment when he has no knowledge of the developmental anatomy of his embryo? It is still most important that a researcher in this field be master of this microanatomy.

When this work was planned, it was necessary to decide which method should be followed. Particularly this question arose: Should the pictures be photographed or could they be hand drawn? The first possibility was the easier way, because only labeling was needed after the photographs were taken. Certainly the pictures were perfect copies of the slide. The second method was more laborious, and the hand drawings could never be so nearly perfect as the photographs. But the photographic method had at least two disadvantages. First, the photographs lose their sharp outline in the printing. Many times the students are confused by the pale figures. Second, the photographs can be focused only at one level of the slide and are not able to show the layers and structures outside of the focused level. For example, if the photograph shows the ventricle of the whole mount, then the atrium and the sinus venosus are not visible. This disadvantage can be eliminated by the drawings, and some detailed structure, unstressed by the camera, can be emphasized. Therefore the decision was made in favor of the hand drawings, even though they involved much more work.

This laboratory atlas is constructed in such a way that it can be used in different kinds of embryology courses. For example, the material in this atlas is sufficient for a two-semester course. However, in most cases the embryology course is only one semester long, hence the instructor can select the material that he wants to teach.

Unfortunately, a certain percentage of students have difficulties in the embryology laboratory. Some have excellent results in comparative anatomy, but have no success in the embryology laboratory. The possible reasons for this are the following:

1. They cannot use the microscope properly.
   a. For instance, they cannot use the correct amount of light to yield the best picture with the most detailed structures at a certain magnification.
   b. They do not use the fine and coarse adjustments properly, particularly in high-power examination and with whole mounts.
2. Of foremost importance is that some students have difficulties in reconstructing the embryo from the two-dimensional slide to the three-dimensional whole

embryo. In comparative anatomy the specimen is three-dimensional and, the structures are palpable and easily visible.

The first two problems can be prevented in the general biology class. The third cannot, but it can be overcome in several ways:

1.   Cross and sagittal sections should be taught in comparative anatomy. This way the student can develop a sense for placing the two-dimensional picture into the whole three-dimensional system.

2.   If the student has not had a comparative anatomy course previously, the instructor can illustrate with a simple analogy, such as the U-shaped tube that can be cut in different planes.

The major goal of this work is to give the student the most help in the embryology laboratory. It is hoped that this atlas will aid students in learning descriptive anatomy, either independently or with the minimal help of their instructors. To fulfill this aim, each figure is fully labeled and the position of the cross sections is indicated on a mid-sagittal section also.

My grateful thanks go to Dr. G. Schreckenberg, whose scrupulous criticisms of this manuscript helped to improve it. I am thankful also to my wife, whose technical assistance, criticism, encouragement, and understanding made this work possible.

Abbreviations

A.	artery
Aa.	arteries
Ant.	anterior
Comm.	common
Dors.	dorsalis
Ext.	external
Inf.	inferior
Int.	internal
Gang.	ganglion
Gl.	gland
L.	left
Lat.	lateral
N.	nerve
Nn.	nerves
Post.	posterior
R.	right
Sag. sec.	sagittal section
Sec.	section
Sup.	superior
Trans.	transverse
Tr.	truncus
V.	vein
Vv.	veins
Vent.	ventral
W.m.	whole mount
x. sec.	cross section
x. 40	magnification; 40 times

Part I

Spermatogenesis

Order: Orthoptera

Grasshopper

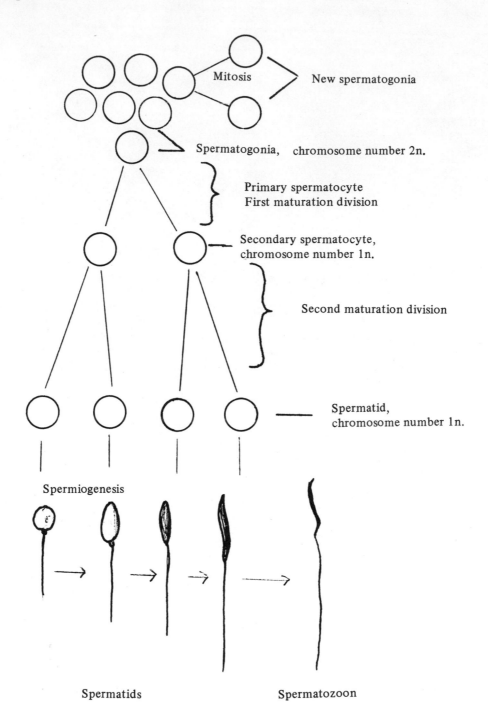

Mitosis → New spermatogonia

Spermatogonia, chromosome number 2n.

Primary spermatocyte
First maturation division

Secondary spermatocyte,
chromosome number 1n.

Second maturation division

Spermatid,
chromosome number 1n.

Spermiogenesis

Spermatids

Spermatozoon

Apex of the testicular lobe

Tunica albuginea

Testicular lobe

Vasa deferens

Ductuli efferentes

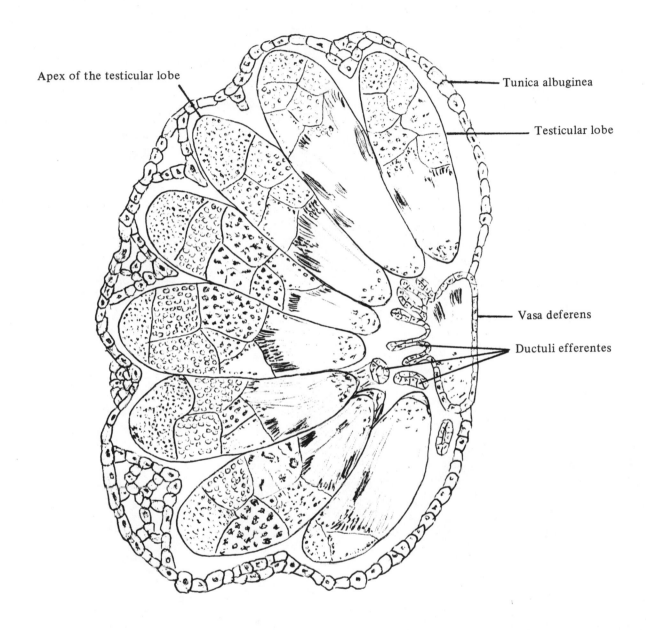

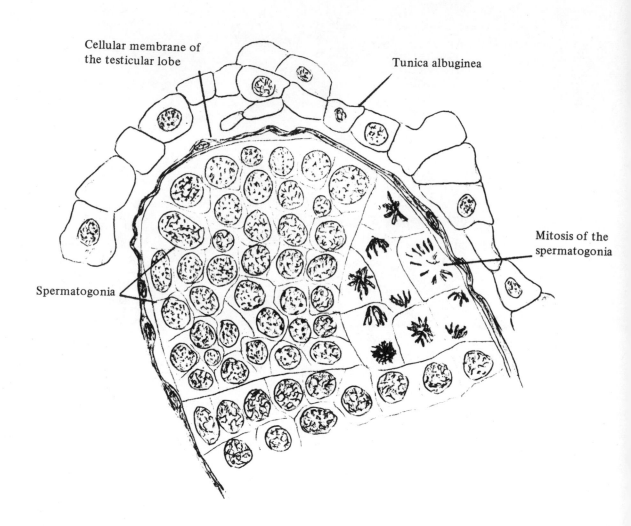

Cellular membrane of
the testicular lobe

Tunica albuginea

Mitosis of the
spermatogonia

Spermatogonia

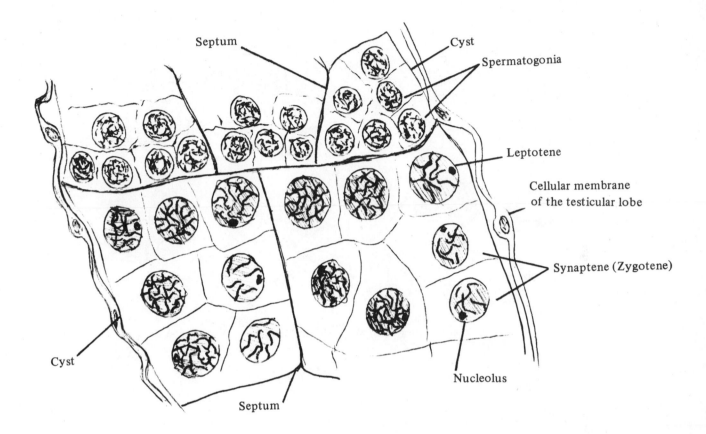

x 1,000

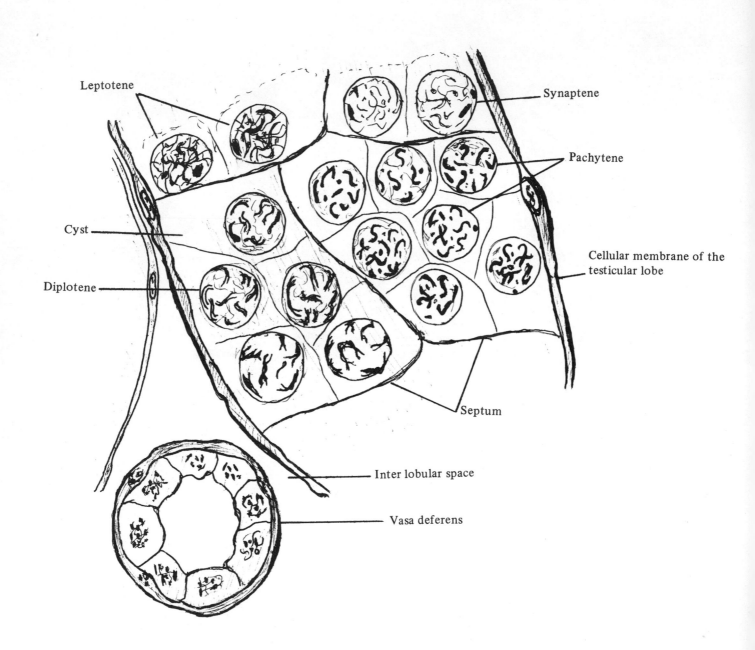

Leptotene

Synaptene

Pachytene

Cyst

Cellular membrane of the
testicular lobe

Diplotene

Septum

Inter lobular space

Vasa deferens

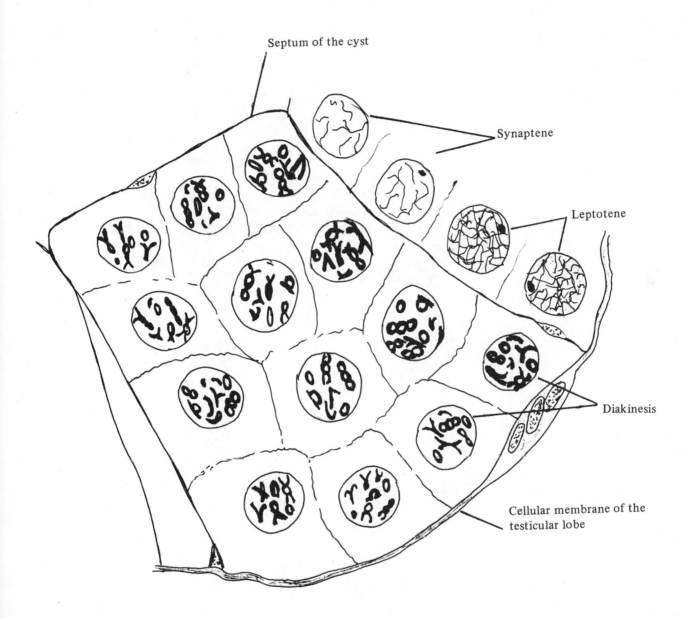

Septum of the cyst

Synaptene

Leptotene

Diakinesis

Cellular membrane of the testicular lobe

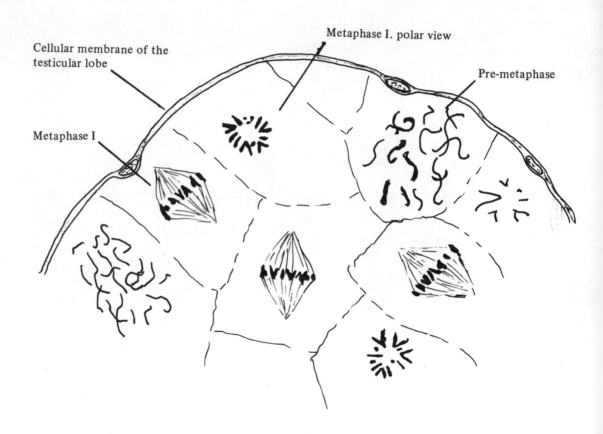

Cellular membrane of the testicular lobe

Metaphase I. polar view

Pre-metaphase

Metaphase I

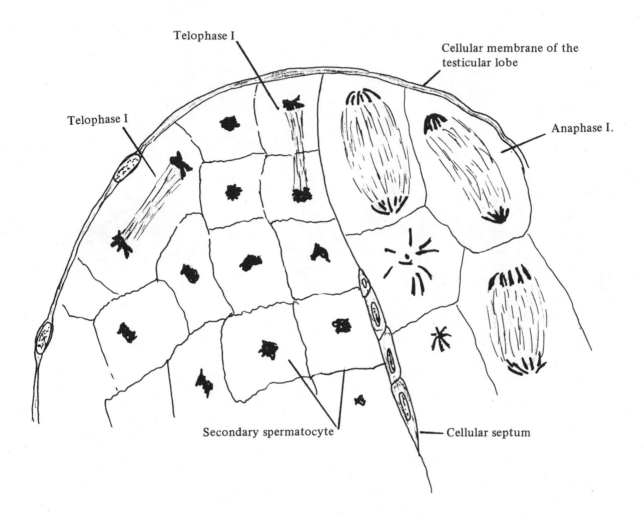

Telophase I

Cellular membrane of the testicular lobe

Telophase I

Anaphase I.

Secondary spermatocyte

Cellular septum

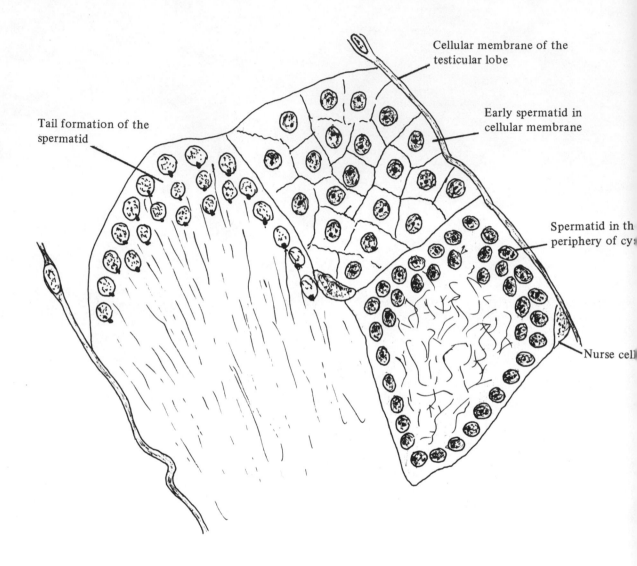

Tail formation of the spermatid

Cellular membrane of the testicular lobe

Early spermatid in cellular membrane

Spermatid in th periphery of cy

Nurse cell

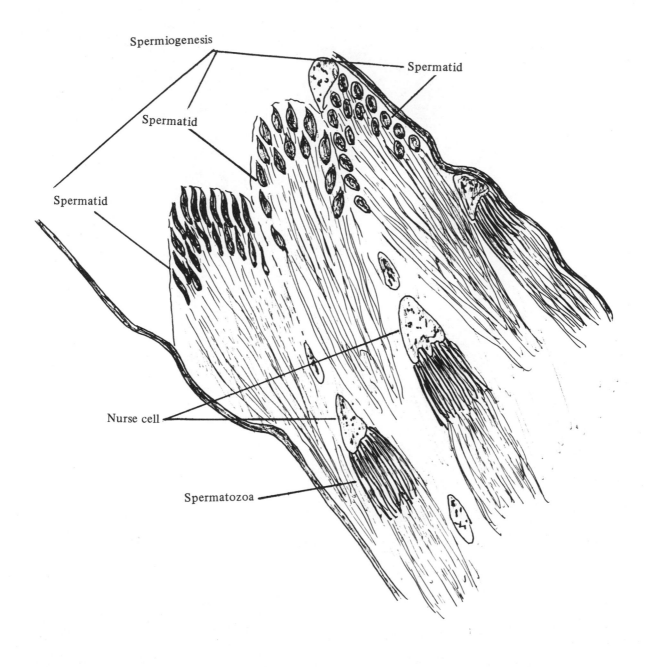

Spermiogenesis

Spermatid

Spermatid

Spermatid

Spermatid

Nurse cell

Spermatozoa

Oogenesis

Order: Rhabditida

Ascaris

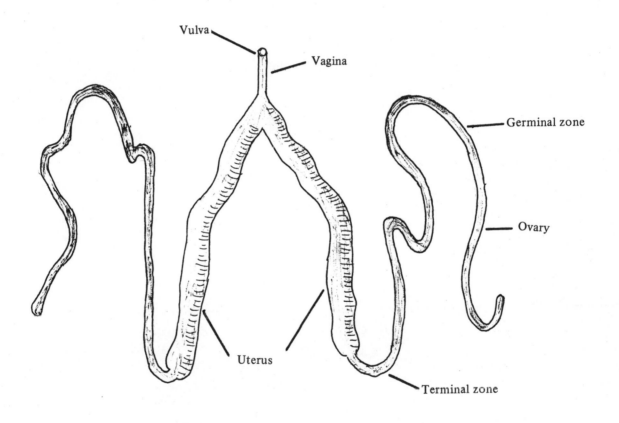

The female genital tract

Diagram

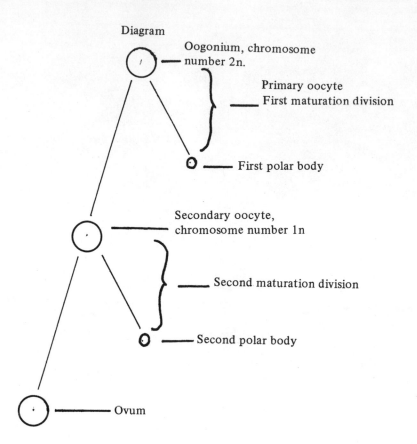

Oogonium, chromosome number 2n.

Primary oocyte
First maturation division

First polar body

Secondary oocyte, chromosome number 1n

Second maturation division

Second polar body

Ovum

Rhabditida: Ascaris

x 1,000

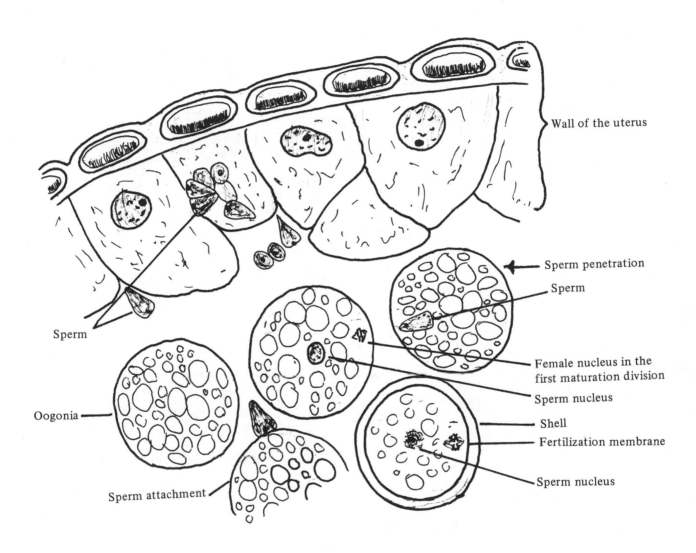

Wall of the uterus

Sperm penetration

Sperm

Female nucleus in the first maturation division

Sperm nucleus

Sperm

Shell

Fertilization membrane

Sperm nucleus

Oogonia

Sperm attachment

x 1,000

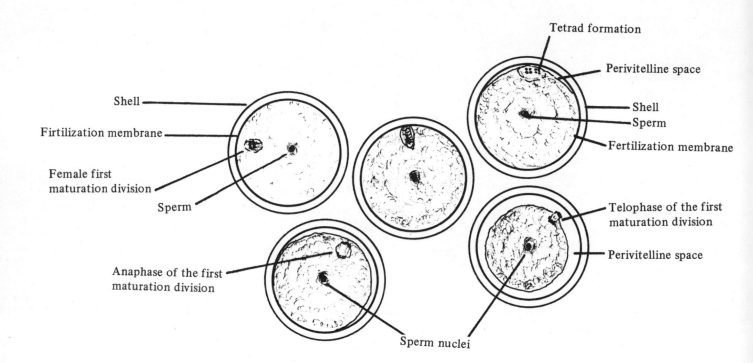

x 1,000

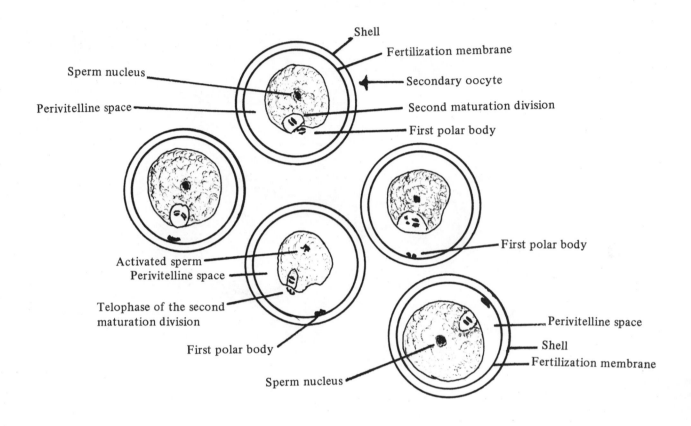

Shell
Fertilization membrane
Sperm nucleus
Secondary oocyte
Perivitelline space
Second maturation division
First polar body

Activated sperm
Perivitelline space
First polar body

Telophase of the second
maturation division

First polar body

Perivitelline space
Shell
Fertilization membrane

Sperm nucleus

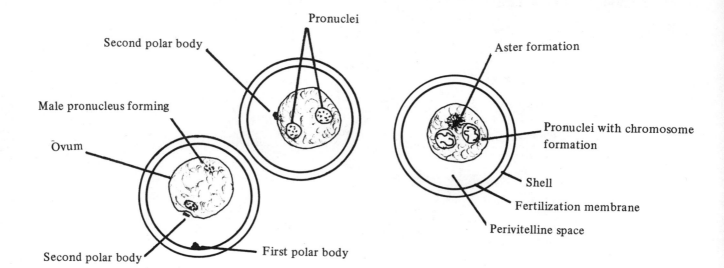

Pronuclei

Second polar body

Male pronucleus forming

Ovum

Second polar body

First polar body

Aster formation

Pronuclei with chromosome formation

Shell

Fertilization membrane

Perivitelline space

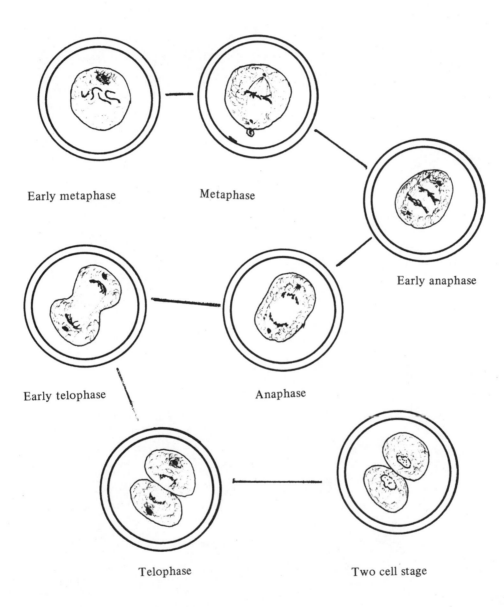

Early metaphase

Metaphase

Early anaphase

Early telophase

Anaphase

Telophase

Two cell stage

The first cleavage — mitosis

Four cell stage: Bilateral cleavage

Early Development of the Starfish

Class: Asteroidea

Astarias vulgaris

Development of the Starfish.   Astarias vulgaris

x 100

Vitelline membrane
Nucleus
Nucleolus

Oogonium

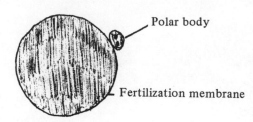

Polar body
Fertilization membrane

Polar body

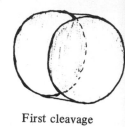

First cleavage
2 cell stage

Third cleavage
8 cell stage

Second cleavage
4 cell stage

Fourth cleavage
16 cell stage
Morula

Fertilization membrane
Blastoderm
Blastocoele

Early blastula

34

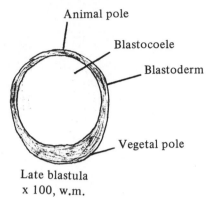

Animal pole

Blastocoele

Blastoderm

Vegetal pole

Late blastula
x 100, w.m.

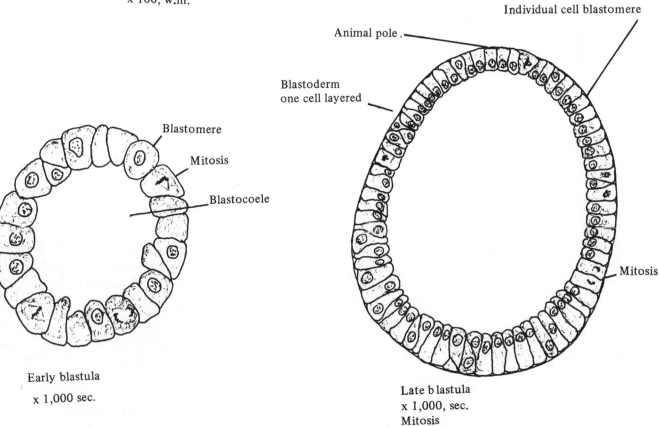

Blastomere

Mitosis

Blastocoele

Early blastula
x 1,000 sec.

Individual cell blastomere

Animal pole

Blastoderm
one cell layered

Mitosis

Late blastula
x 1,000, sec.
Mitosis

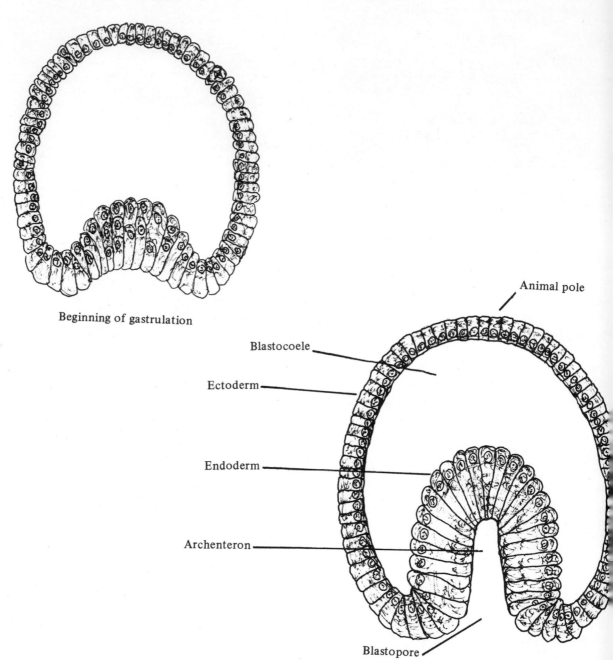

Beginning of gastrulation

Animal pole

Blastocoele

Ectoderm

Endoderm

Archenteron

Blastopore

Early gastrula

Development of the Starfish   Astarias vulgaris

x 1,000

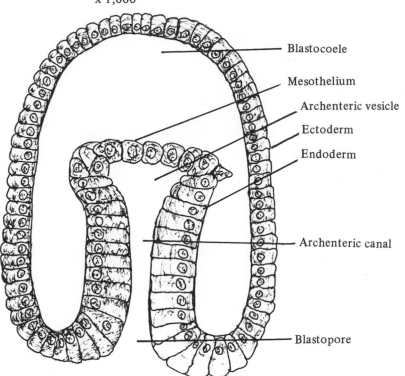

Blastocoele

Mesothelium

Archenteric vesicle

Ectoderm

Endoderm

Archenteric canal

Blastopore

Late gastrula, sec.

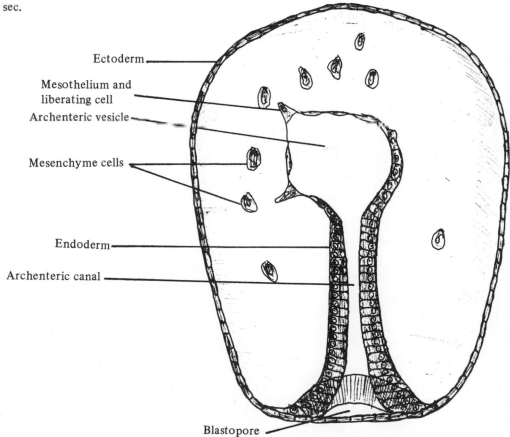

Ectoderm

Mesothelium and
liberating cell

Archenteric vesicle

Mesenchyme cells

Endoderm

Archenteric canal

Blastopore

37

Mesenchyme cell formation
x 400: w.m.

Development of Amphioxus

Subphylum: Cephalochordata

Branchiostoma lanceolatum

Development of Amphioxus:
x 400 w.m.

Branchiostoma lanceolatum

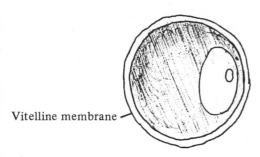

Vitelline membrane

Unfertilized egg

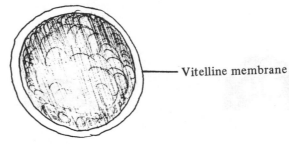

Vitelline membrane

Uncleaved egg

Fertilization membrane

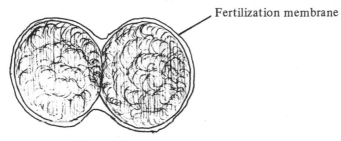

First cleavage
2 cell stage

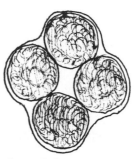

Second cleavage
4 cell stage

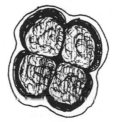

Third cleavage
8 cell stage
Polar view

Fourth cleavage
16 cell stage
Polar view

Development of Amphioxus:

x 400 w.m.

Branchiostoma lanceolatum

Morula

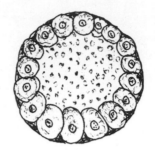

Early blastula

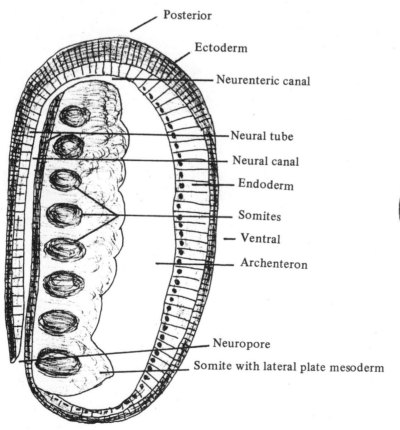

Posterior

Ectoderm

Neurenteric canal

Neural tube

Neural canal

Endoderm

Somites

Ventral

Archenteron

Neuropore

Somite with lateral plate mesoderm

Neural tube

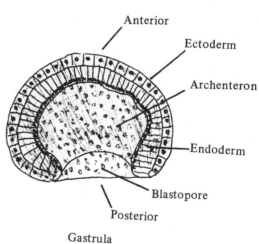

Anterior

Ectoderm

Archenteron

Endoderm

Blastopore

Posterior

Gastrula

Development of Amphioxus:

x 400, w.m.

Branchiostoma lanceolatum

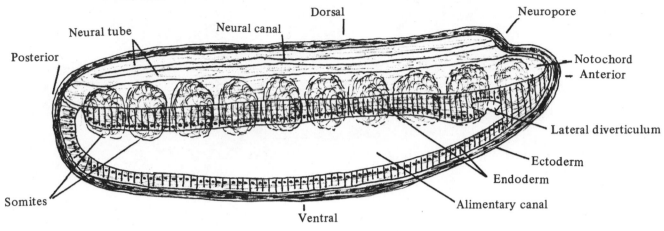

Late embryo

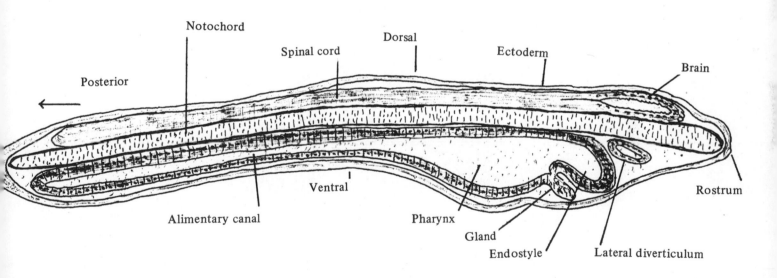

Early larva

Development of Amphioxus:

x 400, w.m.

Branchiostoma lanceolatum

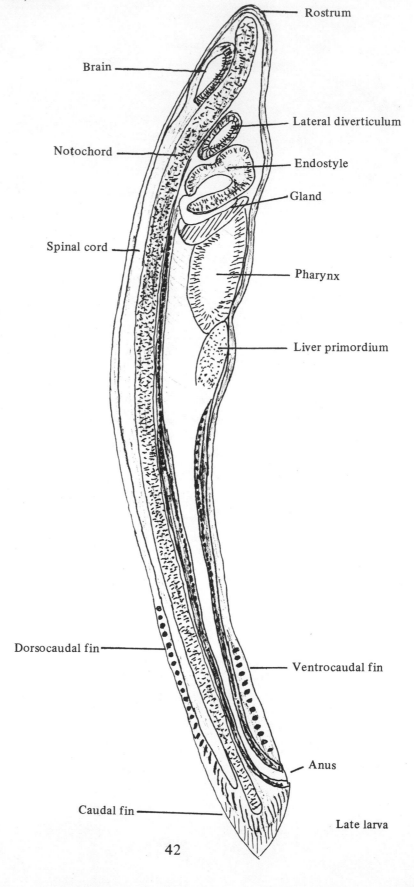

Rostrum

Brain

Lateral diverticulum

Notochord

Endostyle

Gland

Spinal cord

Pharynx

Liver primordium

Dorsocaudal fin

Ventrocaudal fin

Anus

Caudal fin

Late larva

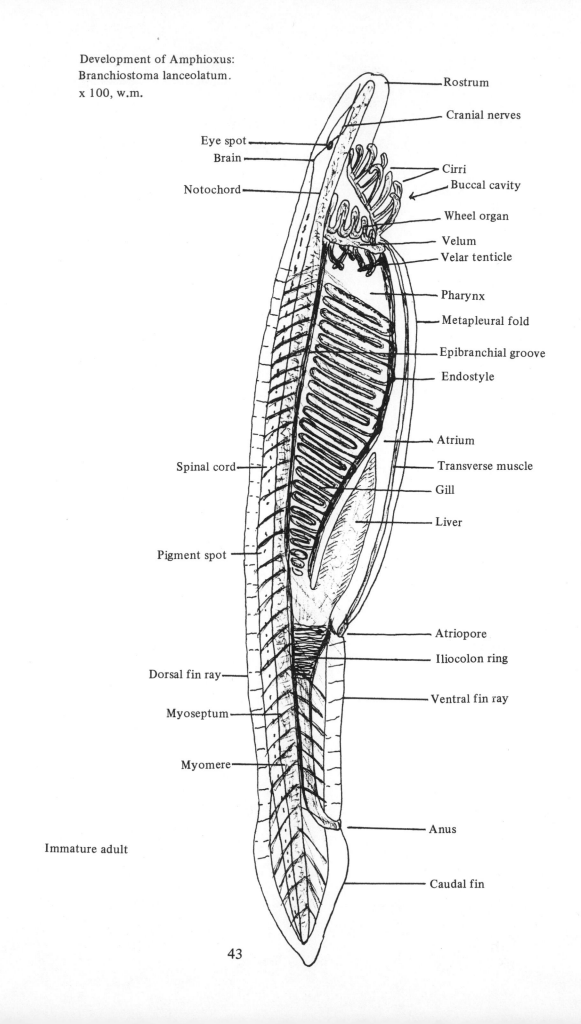

Development of Amphioxus:
Branchiostoma lanceolatum.
x 100, w.m.

Rostrum

Cranial nerves

Eye spot

Brain

Notochord

Cirri

Buccal cavity

Wheel organ

Velum

Velar tenticle

Pharynx

Metapleural fold

Epibranchial groove

Endostyle

Atrium

Transverse muscle

Spinal cord

Gill

Liver

Pigment spot

Atriopore

Iliocolon ring

Dorsal fin ray

Ventral fin ray

Myoseptum

Myomere

Anus

Immature adult

Caudal fin

43

Development of Whitefish

Tanichthys albonubes

Whitefish development 1

x 1.000, sec.

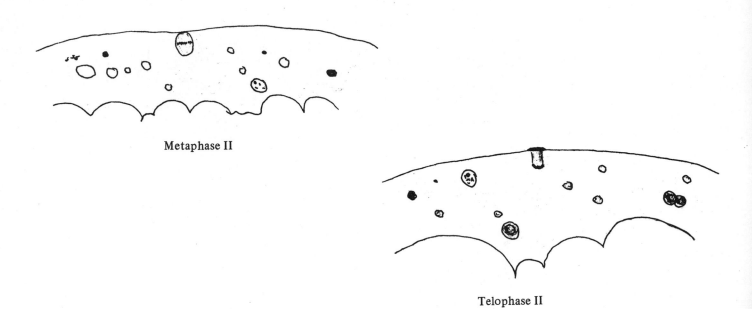

Metaphase II

Telophase II

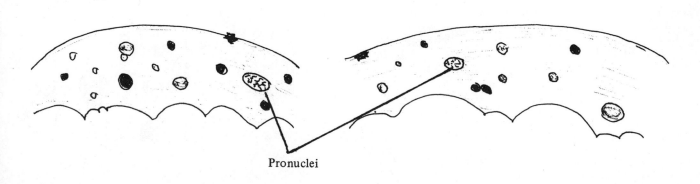

Pronuclei

Oogenesis

Whitefish development 2

x 100, sec.

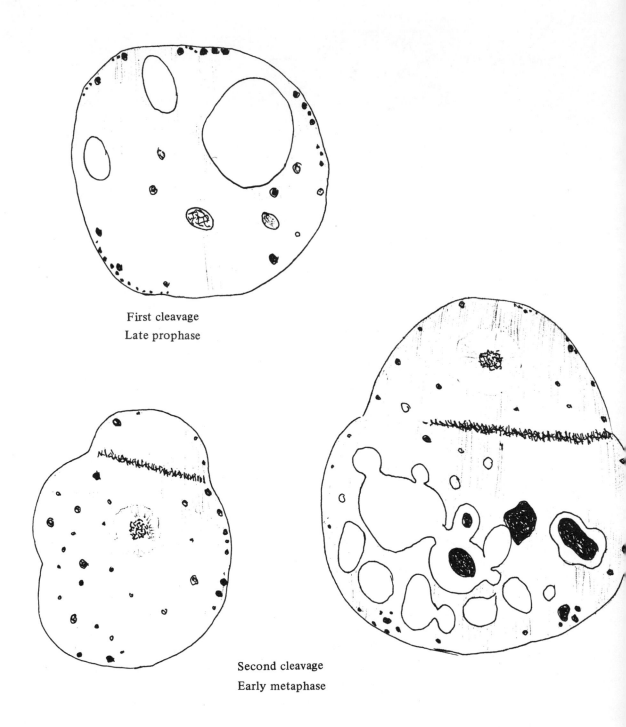

First cleavage
Late prophase

Second cleavage
Early metaphase

Whitefish development 3

x 100, sec.

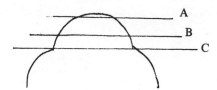

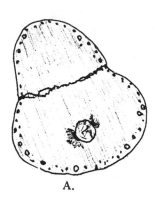

A.

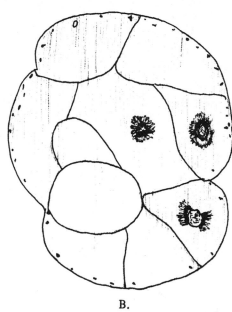

B.

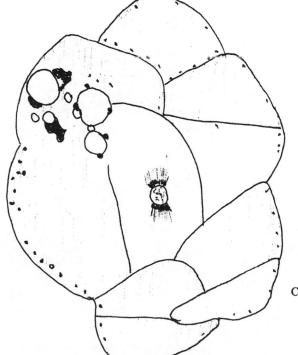

C.

Early cleavage
Early prophase

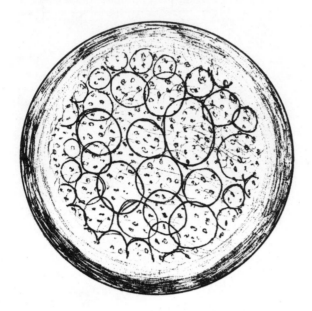

Gastrula
Top view

x 100, w.m.

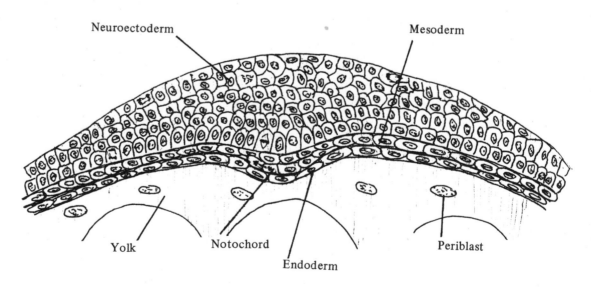

Gastrula
x 400, sec.

Whitefish development 5

Neurulation

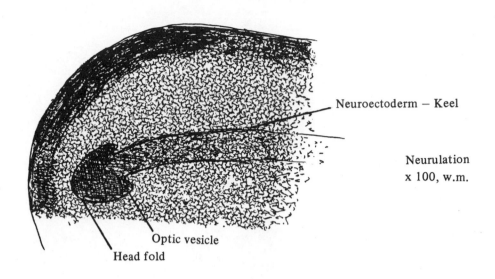

Neuroectoderm – Keel

Neurulation
x 100, w.m.

Optic vesicle

Head fold

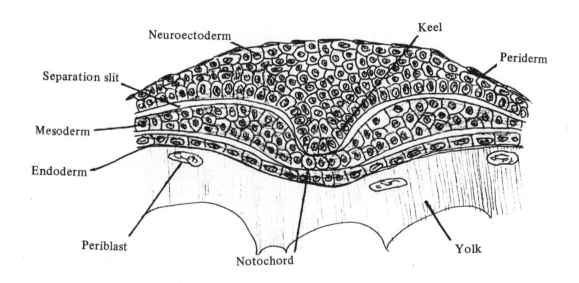

Neuroectoderm

Keel

Separation slit

Periderm

Mesoderm

Endoderm

Periblast

Notochord

Yolk

Neurulation
x 400, x. sec.

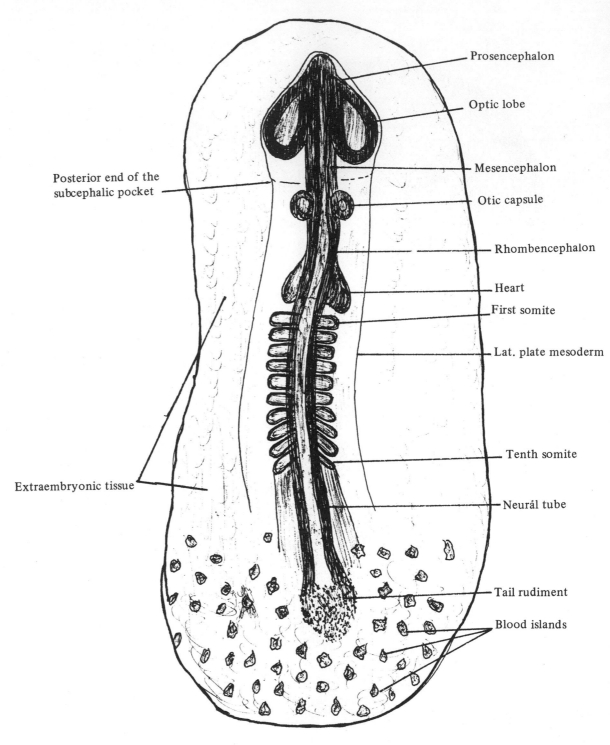

Prosencephalon

Optic lobe

Mesencephalon

Otic capsule

Rhombencephalon

Heart

First somite

Lat. plate mesoderm

Tenth somite

Neurál tube

Posterior end of the
subcephalic pocket

Tail rudiment

Blood islands

Extraembryonic tissue

10 somite stage

10 somite stage

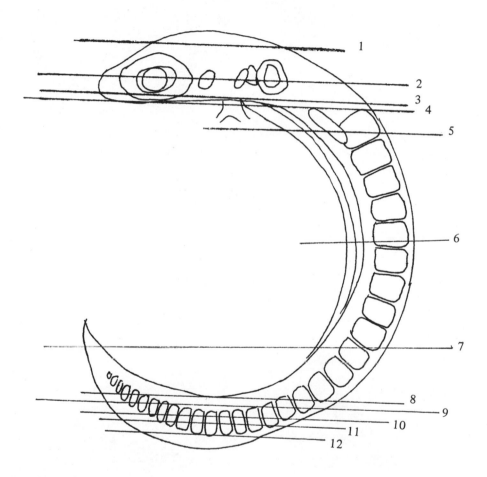

1

2

3

4

5

6

7

8

9

10

11

12

Section planes

Whitefish development 8

x 100, x. sec.

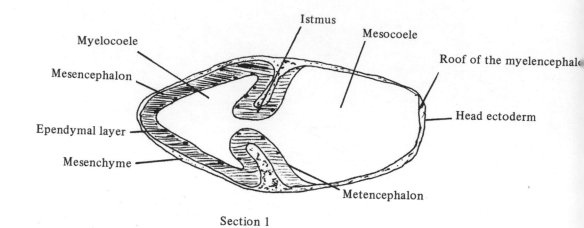

Istmus
Mesocoele
Myelocoele
Roof of the myelencephalo
Mesencephalon
Head ectoderm
Ependymal layer
Mesenchyme
Metencephalon

Section 1

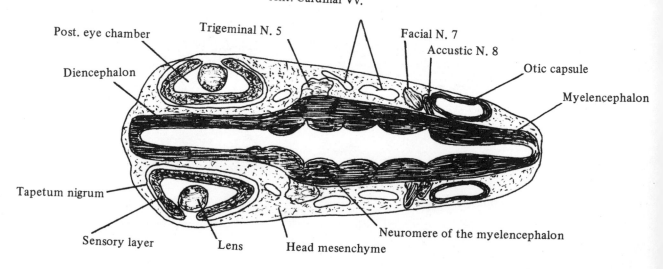

Ant. Cardinal Vv.
Post. eye chamber
Trigeminal N. 5
Facial N. 7
Accustic N. 8
Diencephalon
Otic capsule
Myelencephalon
Tapetum nigrum
Sensory layer
Lens
Head mesenchyme
Neuromere of the myelencephalon

Section 2

Whitefish development 9

x 100, x. sec.

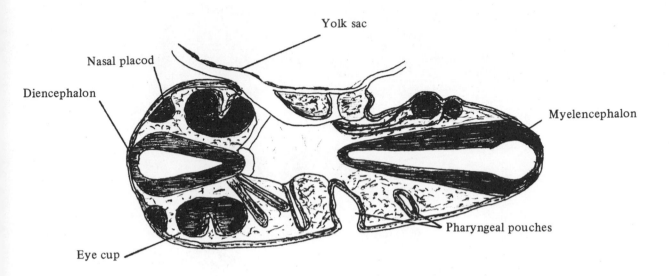

Yolk sac

Nasal placod

Diencephalon

Myelencephalon

Eye cup

Pharyngeal pouches

Section 3

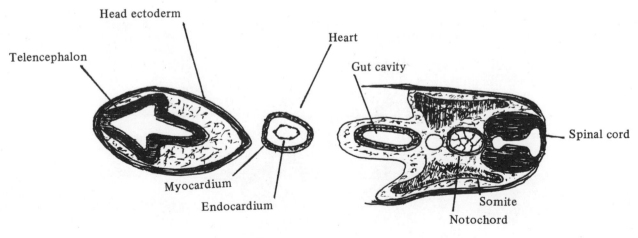

Head ectoderm

Heart

Gut cavity

Telencephalon

Spinal cord

Myocardium

Endocardium

Somite

Notochord

Section 4

Whitefish development 10

x 100, x. sec.

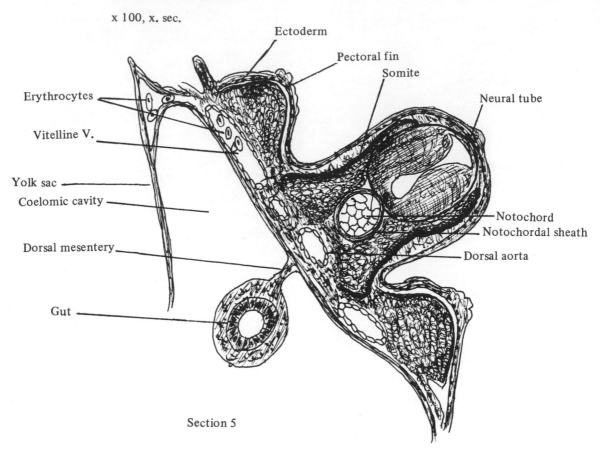

Ectoderm

Pectoral fin

Somite

Neural tube

Erythrocytes

Vitelline V.

Yolk sac

Coelomic cavity

Dorsal mesentery

Gut

Notochord

Notochordal sheath

Dorsal aorta

Section 5

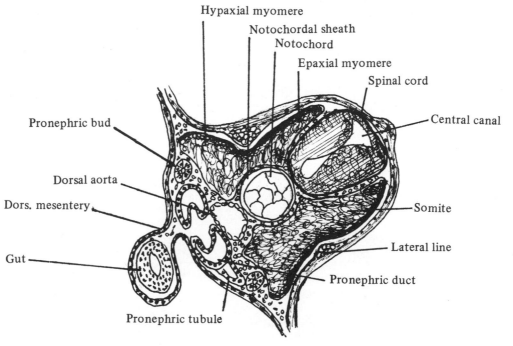

Hypaxial myomere

Notochordal sheath

Notochord

Epaxial myomere

Spinal cord

Pronephric bud

Central canal

Dorsal aorta

Dors. mesentery

Gut

Somite

Lateral line

Pronephric duct

Pronephric tubule

Section 6

Whitefish development 11

x 100, x. sec.

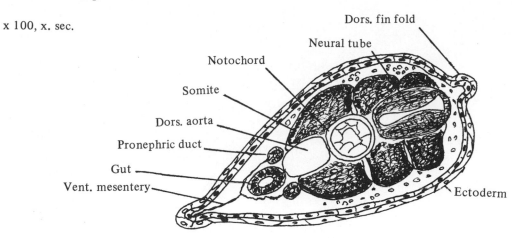

Dors. fin fold

Neural tube

Notochord

Somite

Dors. aorta

Pronephric duct

Gut

Vent. mesentery

Ectoderm

Tail fin fold

Section **7**

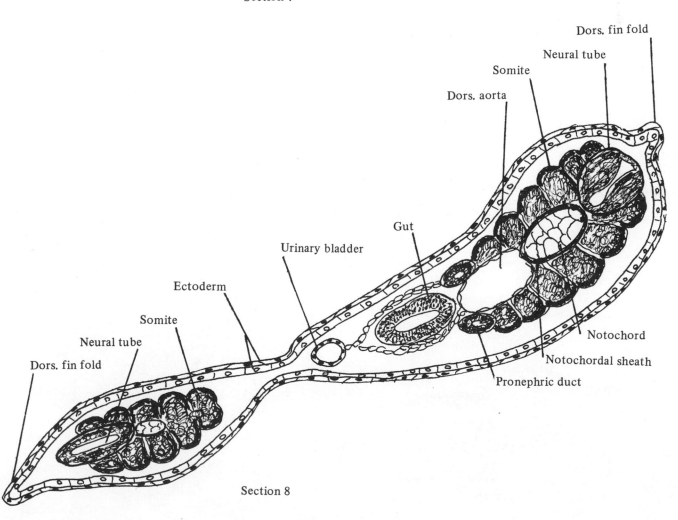

Dors. fin fold

Neural tube

Somite

Dors. aorta

Urinary bladder

Gut

Ectoderm

Somite

Neural tube

Dors. fin fold

Notochord

Notochordal sheath

Pronephric duct

Section 8

Whitefish development 12

x 100, x. sec.

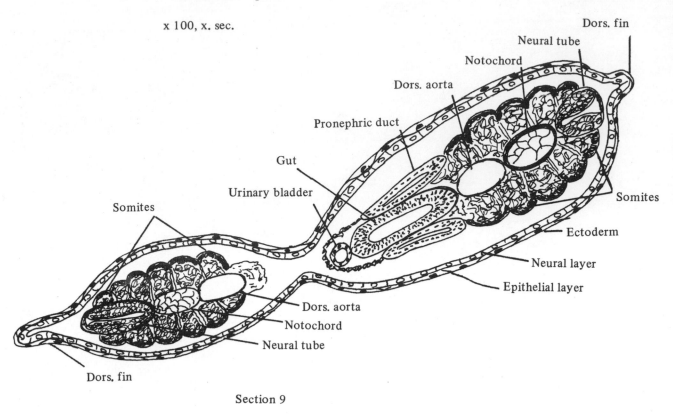

Section 9

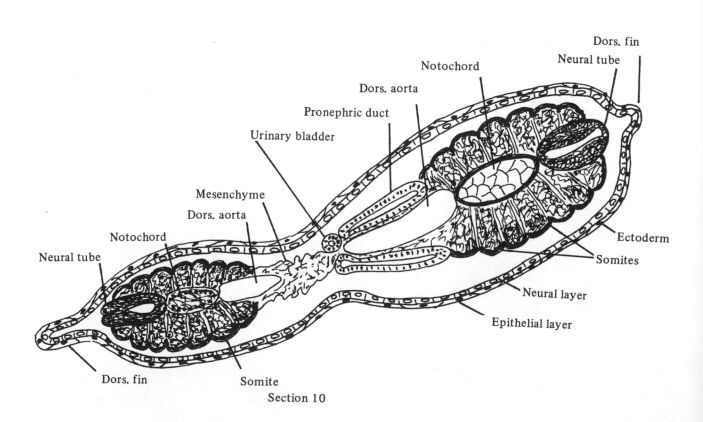

Section 10

Whitefish development 13

x 100, w.m.

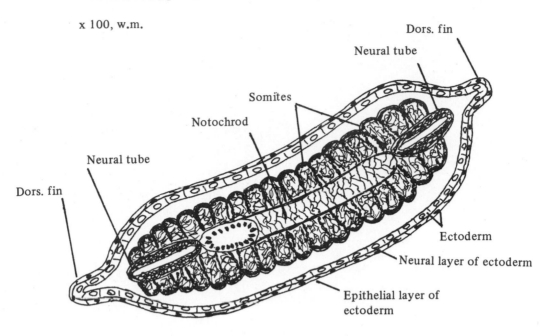

Dors. fin

Neural tube

Somïtes

Notochrod

Neural tube

Dors. fin

Ectoderm

Neural layer of ectoderm

Epithelial layer of ectoderm

Section 11

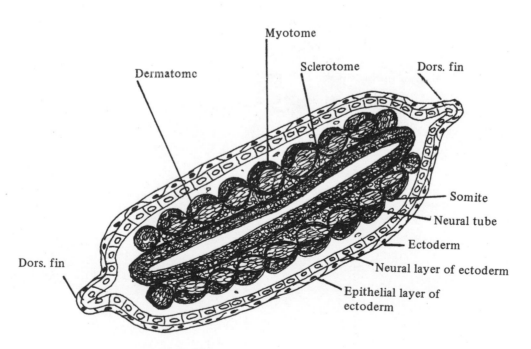

Myotome

Dermatomc

Sclerotome

Dors. fin

Somite

Neural tube

Ectoderm

Neural layer of ectoderm

Epithelial layer of ectoderm

Dors. fin

Section 12

Whitefish development 14

x 400, x. sec.

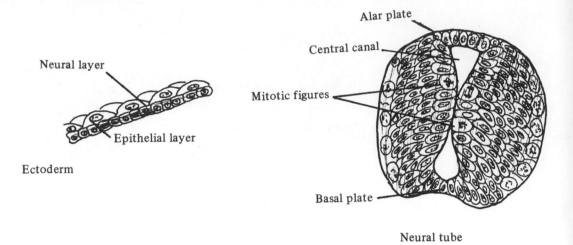

Neural layer

Epithelial layer

Ectoderm

Alar plate

Central canal

Mitotic figures

Basal plate

Neural tube

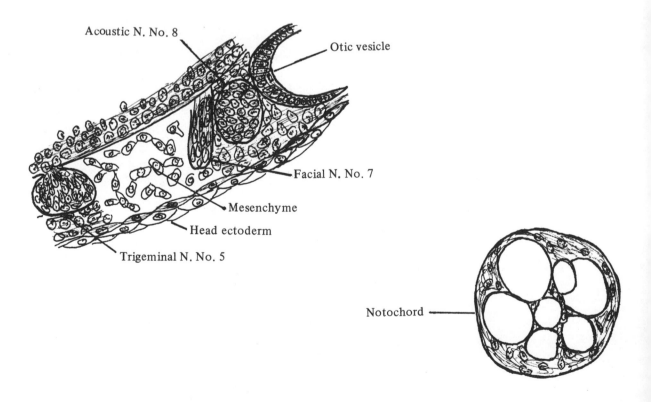

Acoustic N. No. 8

Otic vesicle

Facial N. No. 7

Mesenchyme

Head ectoderm

Trigeminal N. No. 5

Notochord

Frog Development I

Frog Ovary
Frog Testis
Matured Frog Egg
Early Cleavage, Eighth-Cell Stage
Early and Late Blastula
Early and Late Gastrula
Neural Plate

Frog development

Rana pipiens

x 1,000 sec.

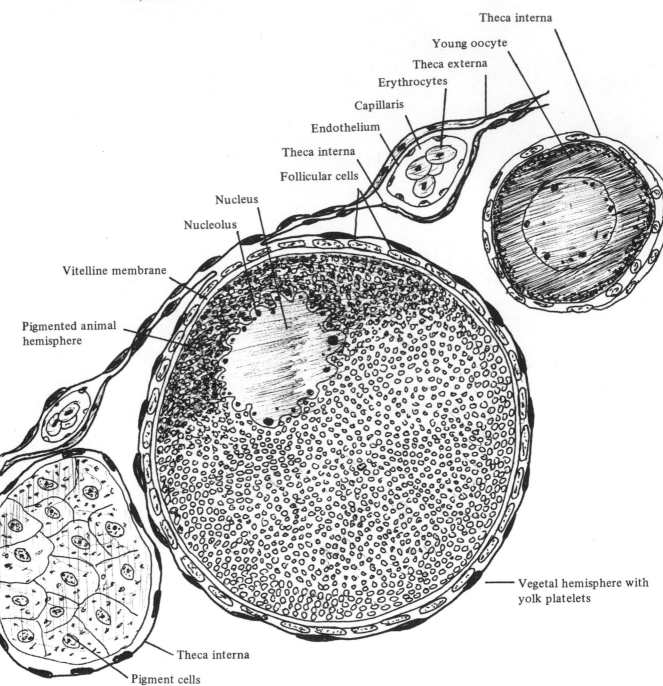

Theca interna

Young oocyte

Theca externa

Erythrocytes

Capillaris

Endothelium

Theca interna

Follicular cells

Nucleus

Nucleolus

Vitelline membrane

Pigmented animal
hemisphere

Vegetal hemisphere with
yolk platelets

Theca interna

Pigment cells

Frog ovary

Frog development   Rana pipiens

x 100, sec.

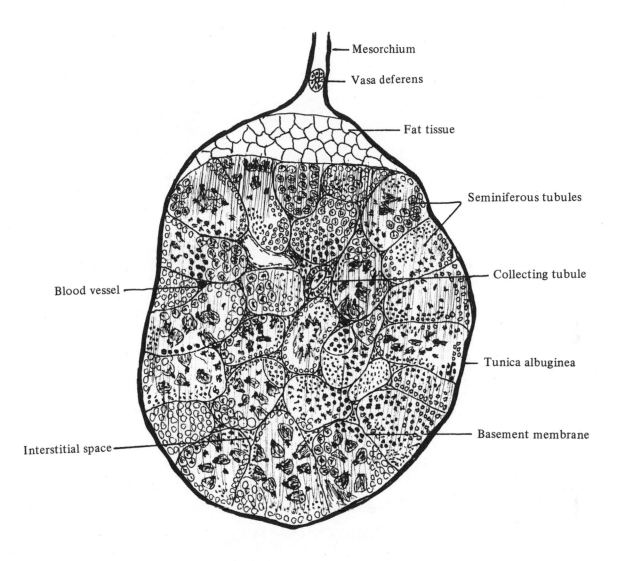

Mesorchium

Vasa deferens

Fat tissue

Seminiferous tubules

Collecting tubule

Blood vessel

Tunica albuginea

Basement membrane

Interstitial space

Frog testis

Frog development   Rana pipiens

x 1,000 sec.

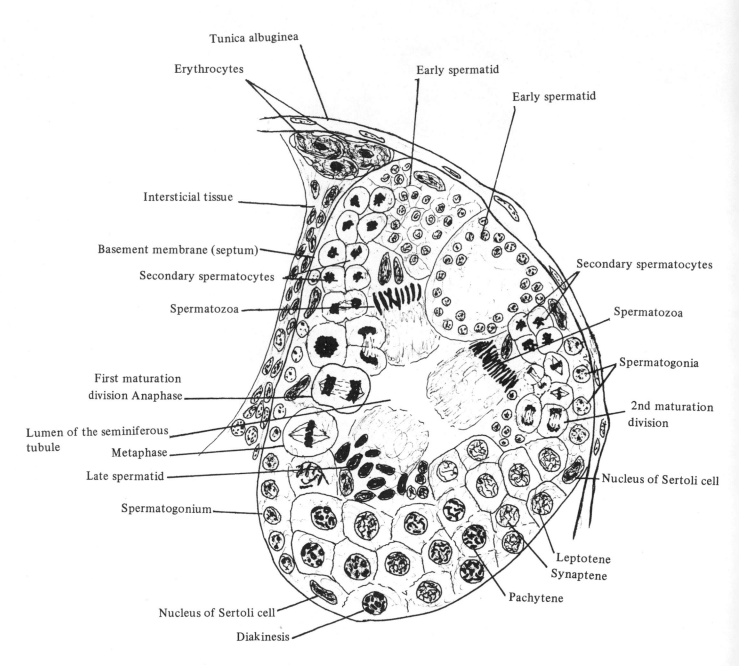

Tunica albuginea

Erythrocytes   Early spermatid

                                                        Early spermatid

Intersticial tissue

Basement membrane (septum)

Secondary spermatocytes   Secondary spermatocytes

Spermatozoa   Spermatozoa

                                                        Spermatogonia

First maturation
division Anaphase   2nd maturation
                                                        division

Lumen of the seminiferous
tubule   Nucleus of Sertoli cell

Metaphase

Late spermatid

Spermatogonium

                                                        Leptotene
                                                        Synaptene

                                                        Pachytene

Nucleus of Sertoli cell

Diakinesis

Seminiferous tubule

Frog development   Rana pipiens

x 1,000 sec.

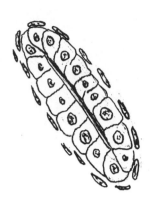

Collecting tubules   Collecting tubules
Spring   Winter

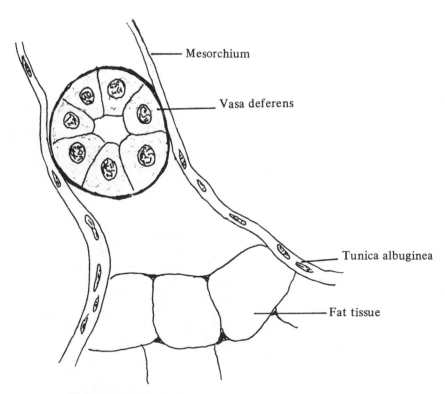

                           — Mesorchium

                          — Vasa deferens

                          — Tunica albuginea

                          — Fat tissue

Mesentery of the testis

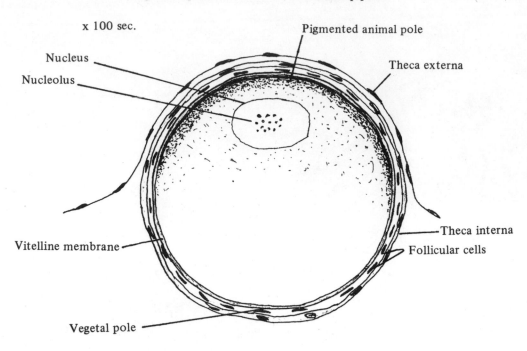

x 100 sec.

Nucleus

Pigmented animal pole

Nucleolus

Theca externa

Theca interna

Vitelline membrane

Follicular cells

Vegetal pole

Mature frog egg

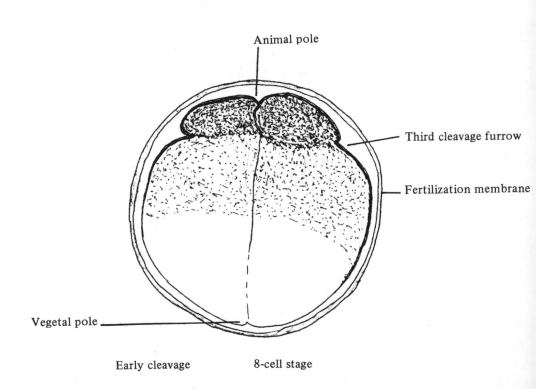

Animal pole

Third cleavage furrow

Fertilization membrane

Vegetal pole

Early cleavage   8-cell stage

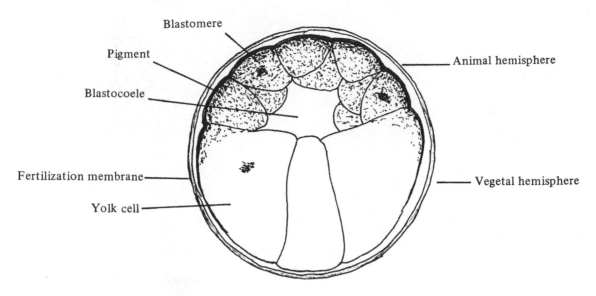

Blastomere

Pigment

Blastocoele

Animal hemisphere

Fertilization membrane

Vegetal hemisphere

Yolk cell

Late cleavage   Early blastula

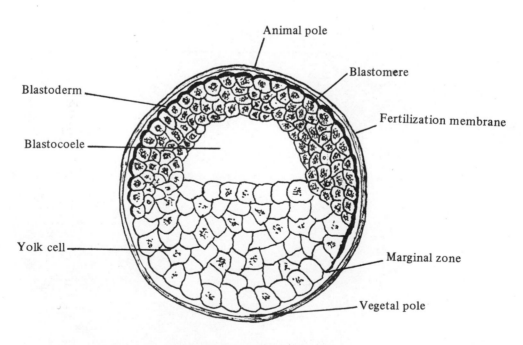

Animal pole

Blastomere

Blastoderm

Fertilization membrane

Blastocoele

Yolk cell

Marginal zone

Vegetal pole

Late blastula

65

Frog development   Rana pipiens

x 100, sec.

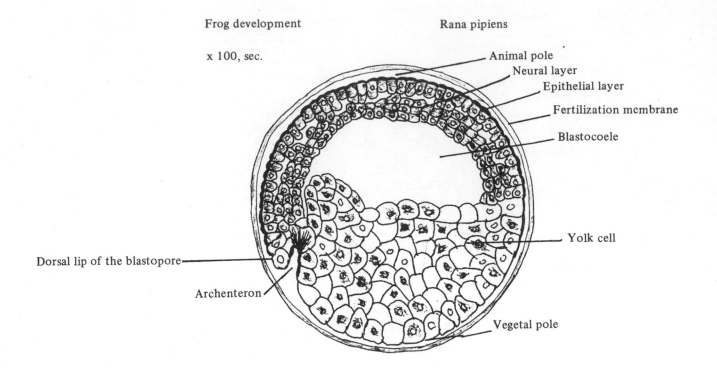

Animal pole
Neural layer
Epithelial layer
Fertilization membrane
Blastocoele

Yolk cell

Dorsal lip of the blastopore

Archenteron

Vegetal pole

Early gastrula

Dorsal   Archenteron

Notochord   Epithelial layer
                                       Neural layer

Endoderm   Gastrular slit

Involuting cells

Dorsal lip of the
blastopore   Completion bridge
Blastopore   Anterior
Posterior
Yolk plug   Blastocoele

Blastopore
Ventral lip of the
blastopore   Yolk cell

Peristomial mesoderm

Ventral

Late gastrula — Yolk plug

Frog development   Rana pipiens

x 100, x. sec.

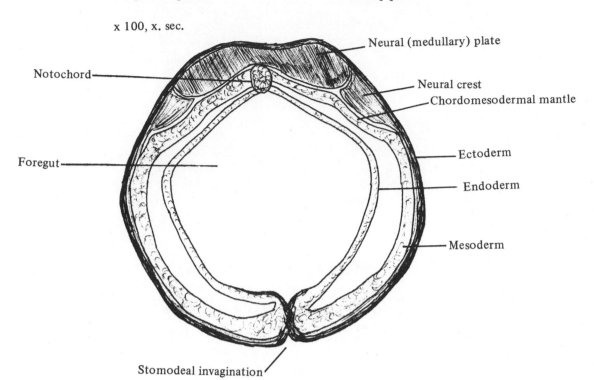

Neural (medullary) plate

Notochord

Neural crest

Chordomesodermal mantle

Ectoderm

Foregut

Endoderm

Mesoderm

Stomodeal invagination

Neural plate — ant. section

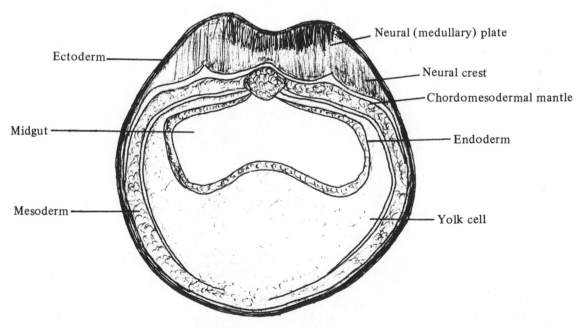

Neural (medullary) plate

Ectoderm

Neural crest

Chordomesodermal mantle

Midgut

Endoderm

Mesoderm

Yolk cell

Neural plate — post. section

Frog Development II

Frog: Neural Fold
3 mm. Frog Embryo

Frog development   Rana pipiens

x 100, median sag. sec.

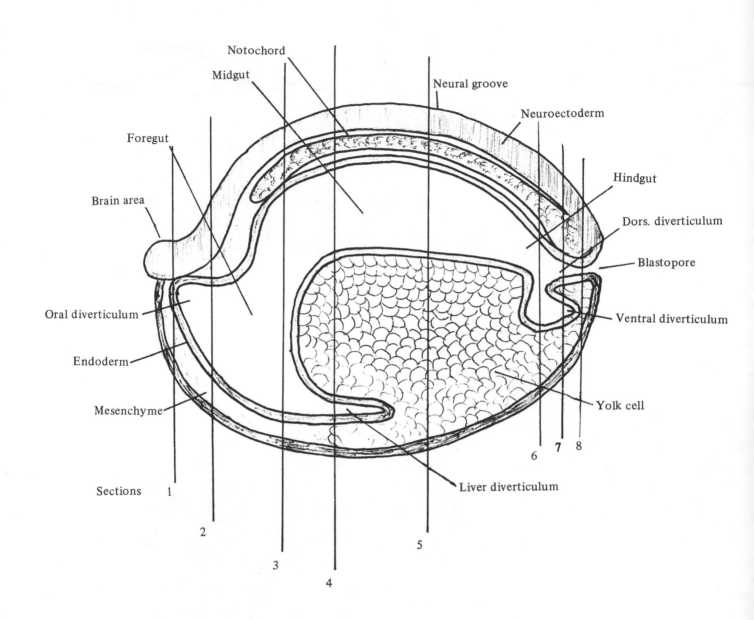

Frog: neural fold

Frog development   Rana pipiens

x 100, x. sec.

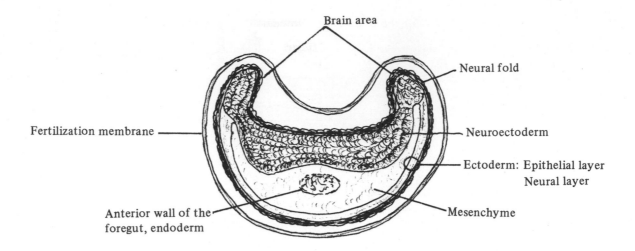

Brain area

Neural fold

Fertilization membrane

Neuroectoderm

Ectoderm: Epithelial layer
Neural layer

Anterior wall of the
foregut, endoderm

Mesenchyme

Section 1

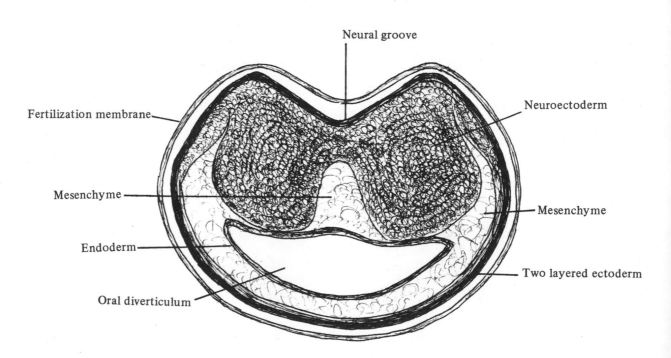

Neural groove

Fertilization membrane

Neuroectoderm

Mesenchyme

Mesenchyme

Endoderm

Two layered ectoderm

Oral diverticulum

Section 2

Frog development   Rana pipiens

x 100, x. sec.

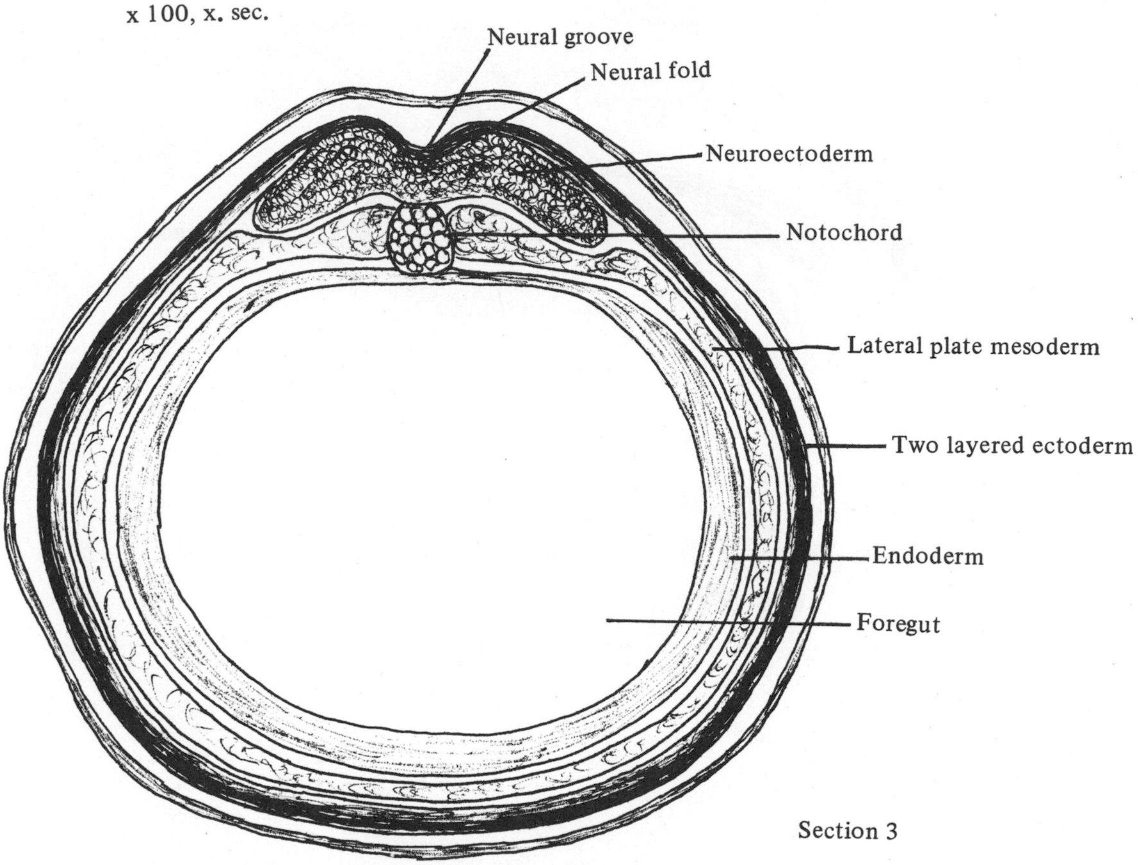

Neural groove
Neural fold
Neuroectoderm
Notochord
Lateral plate mesoderm
Two layered ectoderm
Endoderm
Foregut

Section 3

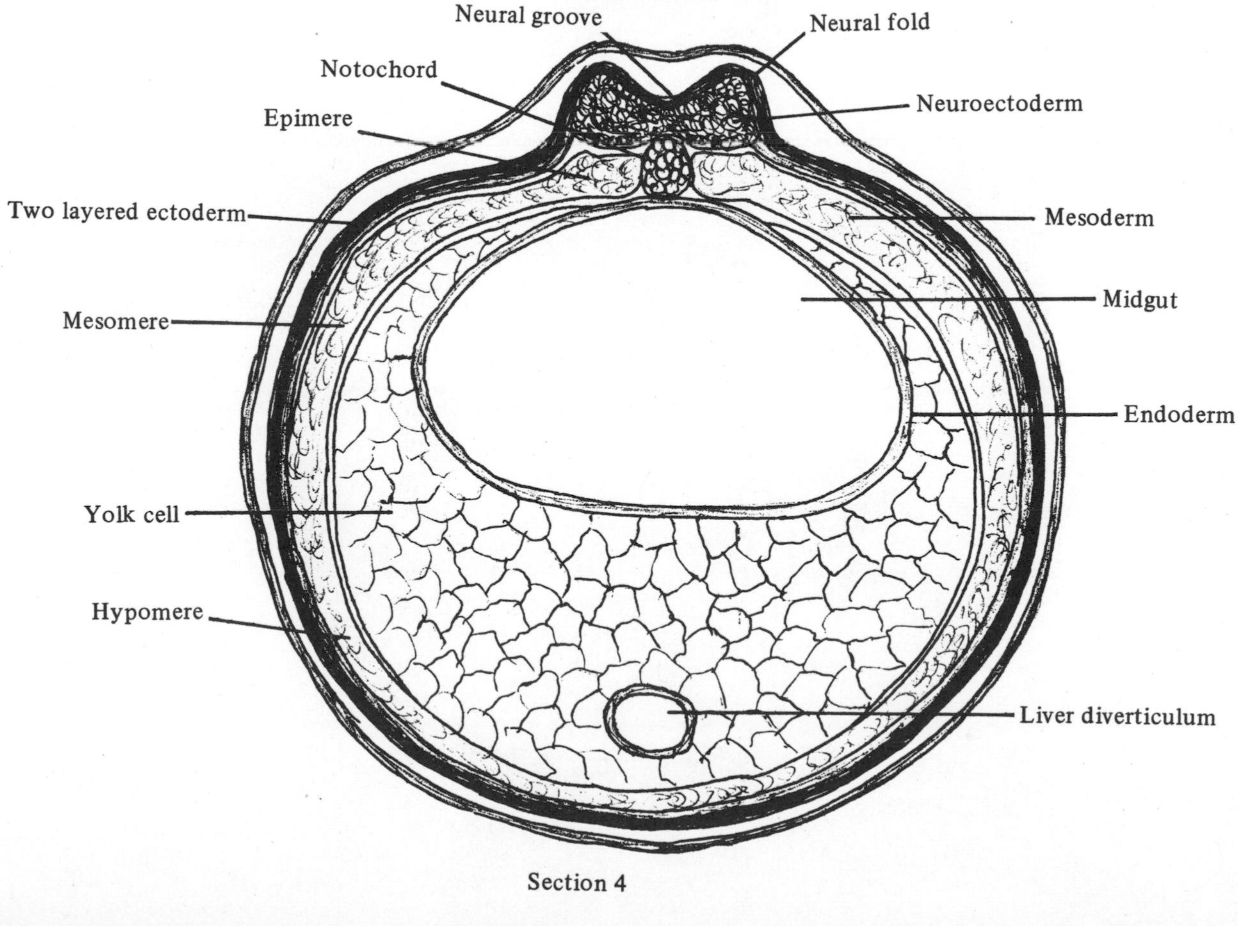

Neural groove   Neural fold
Notochord
Epimere
Neuroectoderm
Two layered ectoderm
Mesoderm
Mesomere
Midgut
Endoderm
Yolk cell
Hypomere
Liver diverticulum

Section 4

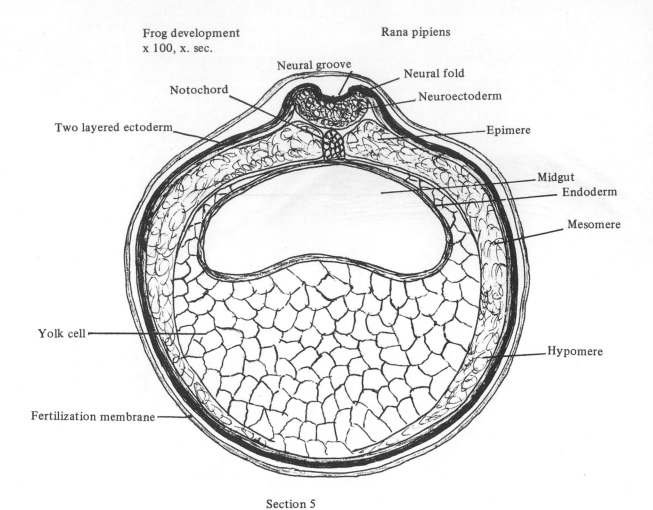

Frog development
x 100, x. sec.

Rana pipiens

Neural groove

Notochord

Neural fold

Neuroectoderm

Two layered ectoderm

Epimere

Midgut

Endoderm

Mesomere

Yolk cell

Hypomere

Fertilization membrane

Section 5

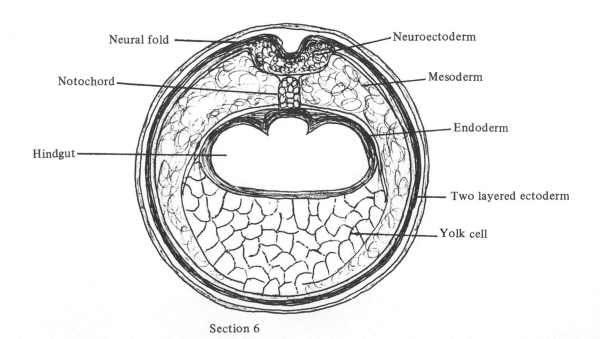

Neural fold

Neuroectoderm

Notochord

Mesoderm

Hindgut

Endoderm

Two layered ectoderm

Yolk cell

Section 6

Frog development   Rana pipiens

x 100, x. sec.

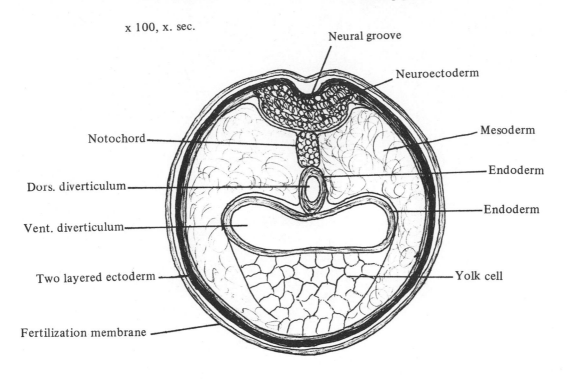

Neural groove

Neuroectoderm

Notochord

Mesoderm

Dors. diverticulum

Endoderm

Vent. diverticulum

Endoderm

Two layered ectoderm

Yolk cell

Fertilization membrane

Section **7**

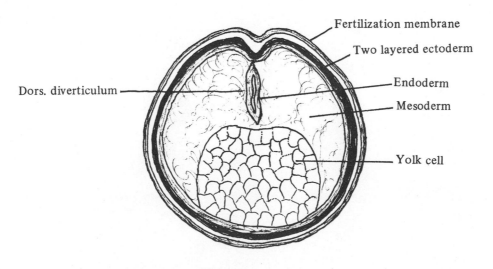

Fertilization membrane

Two layered ectoderm

Dors. diverticulum

Endoderm

Mesoderm

Yolk cell

Section 8

73

Frog development   Rana pipiens

x 100

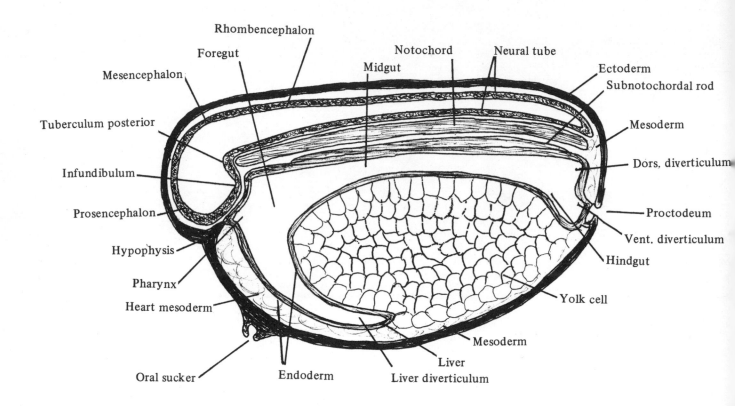

Median sagittal section
3mm. frog embryo

Frog development   Rana pipiens

x 100

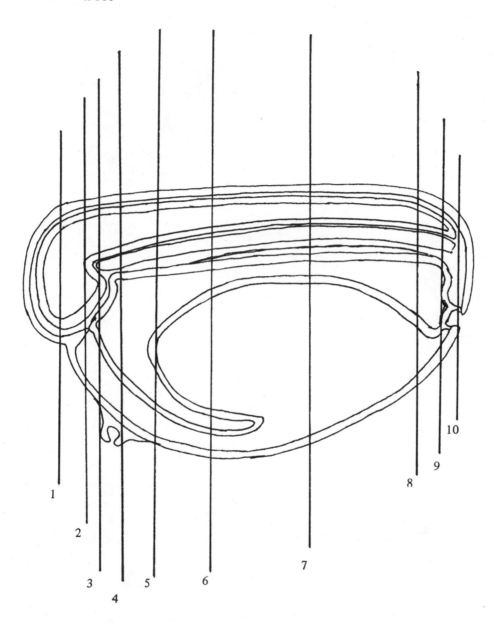

1

2

3   5   6   7

4

Median sagittal section

3 mm. frog embryo

Sections

Frog development

x 100, x. sec.

Rana pipiens

3 mm. embryo

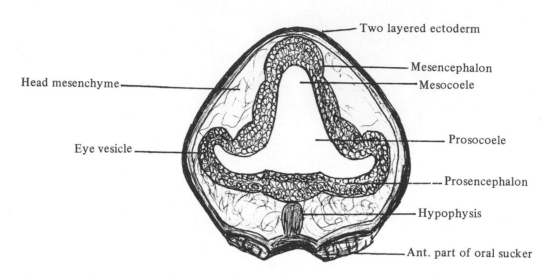

Two layered ectoderm

Mesencephalon

Mesocoele

Head mesenchyme

Prosocoele

Eye vesicle

Prosencephalon

Hypophysis

Ant. part of oral sucker

Section 1

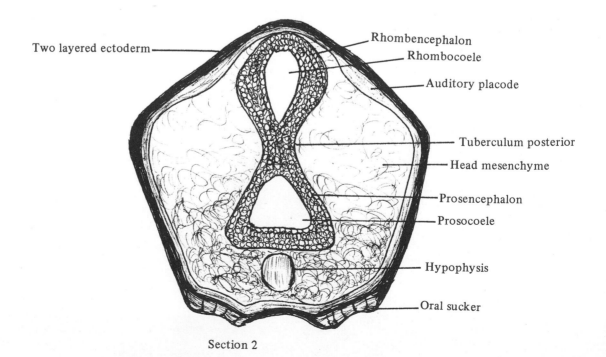

Two layered ectoderm

Rhombencephalon

Rhombocoele

Auditory placode

Tuberculum posterior

Head mesenchyme

Prosencephalon

Prosocoele

Hypophysis

Oral sucker

Section 2

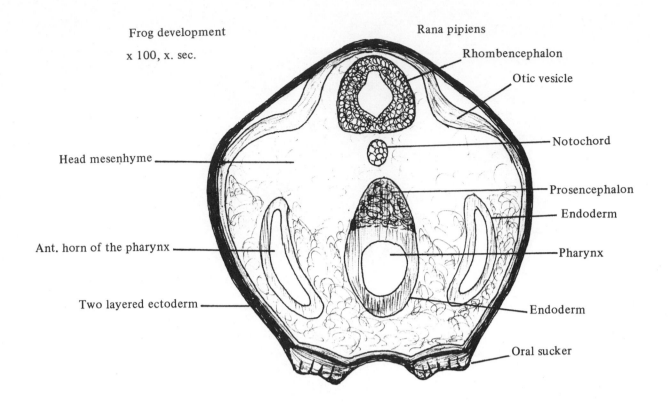

Frog development
x 100, x. sec.

Rana pipiens

Rhombencephalon

Otic vesicle

Notochord

Head mesenhyme

Prosencephalon

Endoderm

Ant. horn of the pharynx

Pharynx

Two layered ectoderm

Endoderm

Oral sucker

Section 3

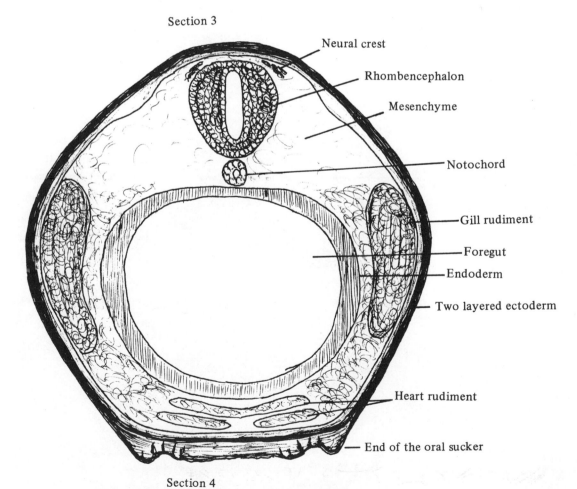

Neural crest

Rhombencephalon

Mesenchyme

Notochord

Gill rudiment

Foregut

Endoderm

Two layered ectoderm

Heart rudiment

End of the oral sucker

Section 4

Frog development   Rana pipiens

x 100, x. sec.   Neural crest

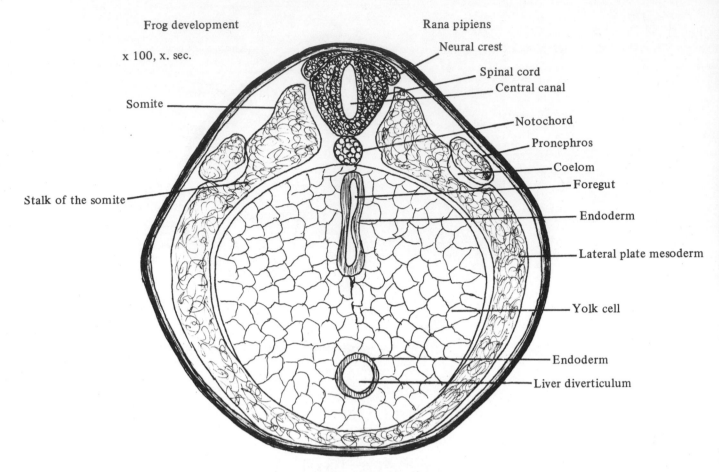

Somite   Spinal cord
                                    Central canal

                                    Notochord

                                    Pronephros

                                    Coelom

Stalk of the somite   Foregut

                                    Endoderm

                                    Lateral plate mesoderm

                                    Yolk cell

                                    Endoderm

                                    Liver diverticulum

Section 5

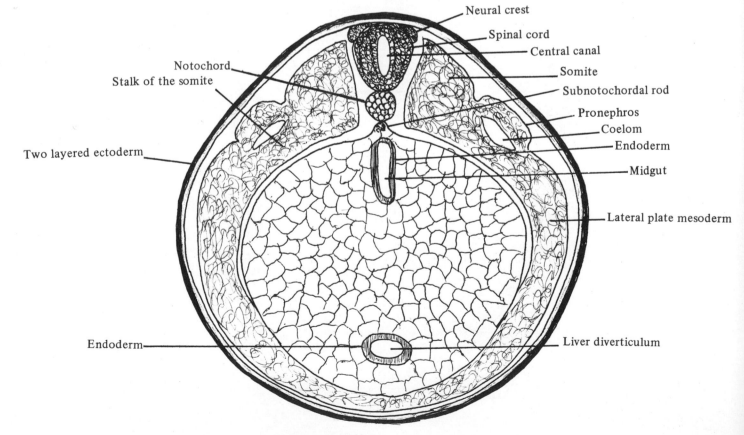

                                    Neural crest

                                    Spinal cord

                                    Central canal

Notochord   Somite

Stalk of the somite   Subnotochordal rod

                                    Pronephros

                                    Coelom

Two layered ectoderm   Endoderm

                                    Midgut

                                    Lateral plate mesoderm

Endoderm   Liver diverticulum

Section 6

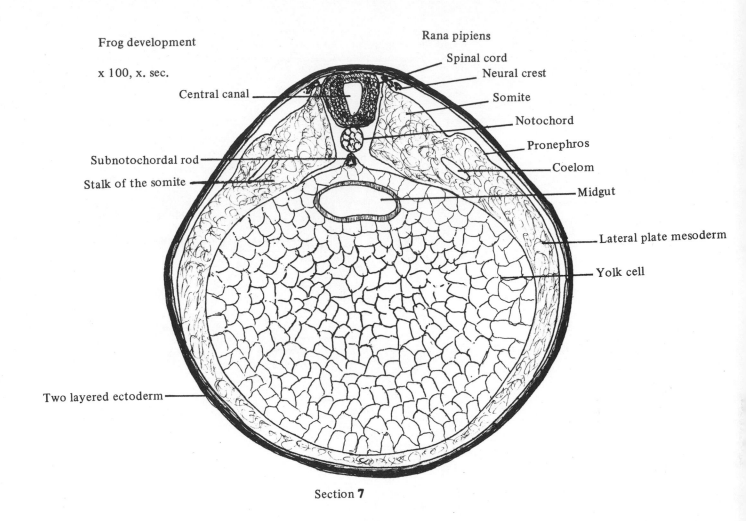

Frog development

x 100, x. sec.

Rana pipiens

Spinal cord

Neural crest

Central canal

Somite

Notochord

Pronephros

Subnotochordal rod

Coelom

Stalk of the somite

Midgut

Lateral plate mesoderm

Yolk cell

Two layered ectoderm

Section **7**

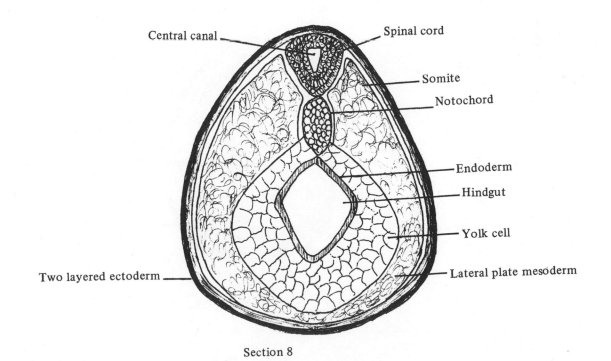

Central canal

Spinal cord

Somite

Notochord

Endoderm

Hindgut

Yolk cell

Two layered ectoderm

Lateral plate mesoderm

Section 8

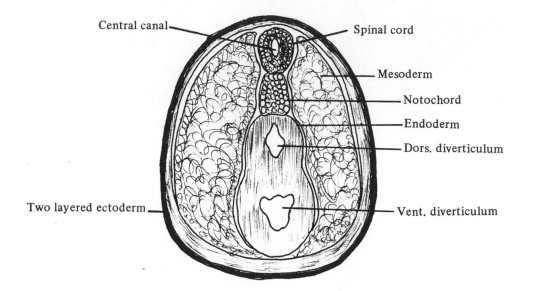

Central canal — Spinal cord

— Mesoderm

— Notochord

— Endoderm

— Dors. diverticulum

Two layered ectoderm — — Vent. diverticulum

Section 9

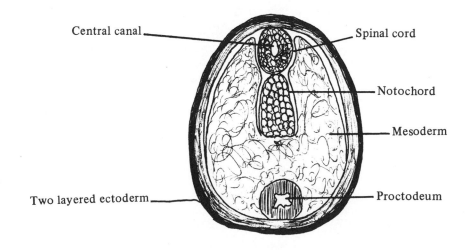

Central canal — Spinal cord

— Notochord

— Mesoderm

Two layered ectoderm — — Proctodeum

Section 10

Frog Development III

4 mm. Frog Embryo
7 mm. Frog Embryo

Frog development   4 mm. embryo

x 100, median sag. sec.

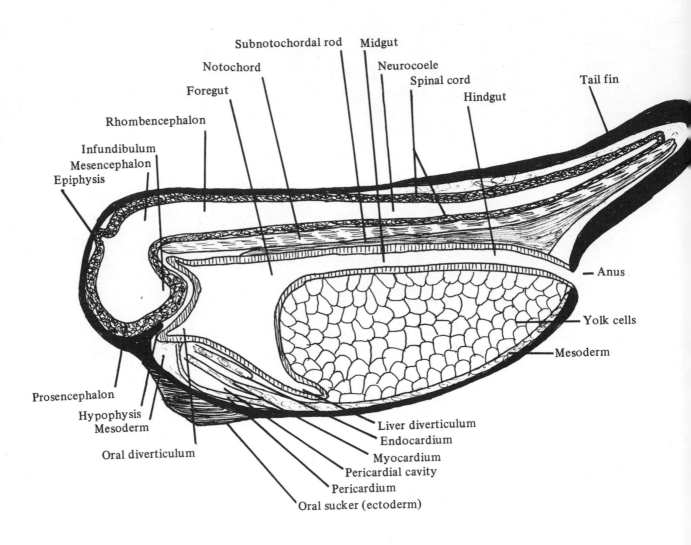

Frog development

4 mm. embryo

x 100, median sag. sec.

SECTIONS:

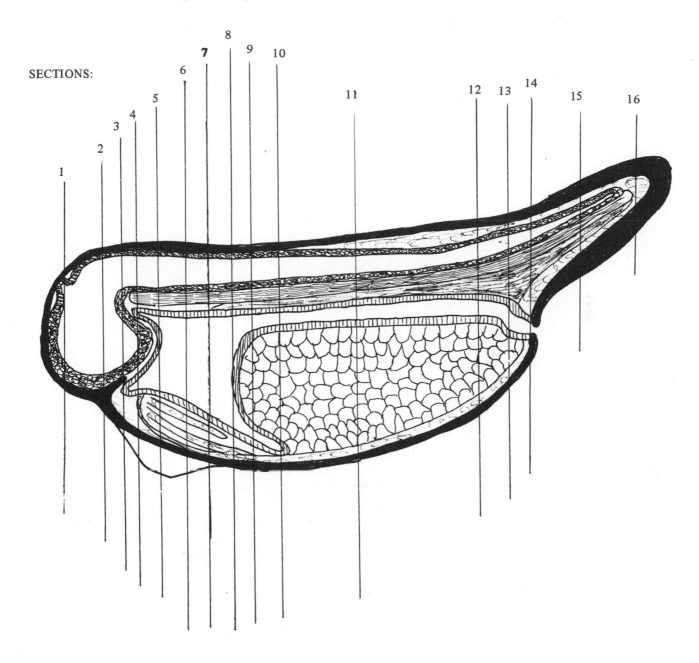

Frog development   4 mm. embryo

x 100, x. sec.

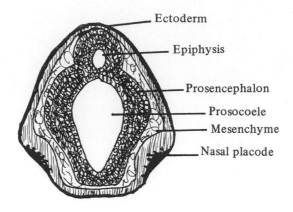

Ectoderm

Epiphysis

Prosencephalon

Prosocoele

Mesenchyme

Nasal placode

Section 1

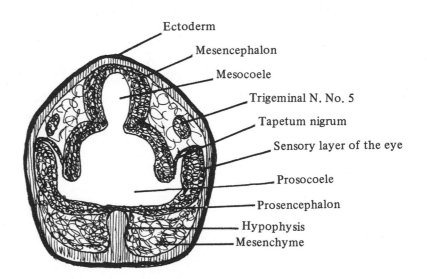

Ectoderm

Mesencephalon

Mesocoele

Trigeminal N. No. 5

Tapetum nigrum

Sensory layer of the eye

Prosocoele

Prosencephalon

Hypophysis

Mesenchyme

Section 2

Frog development 4 mm. embryo

x 100. x. sec.

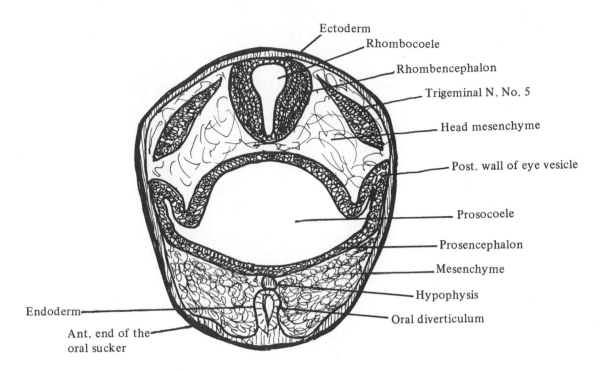

Ectoderm
Rhombocoele
Rhombencephalon
Trigeminal N. No. 5
Head mesenchyme
Post. wall of eye vesicle
Prosocoele
Prosencephalon
Mesenchyme
Hypophysis
Oral diverticulum

Endoderm
Ant. end of the oral sucker

Section 3

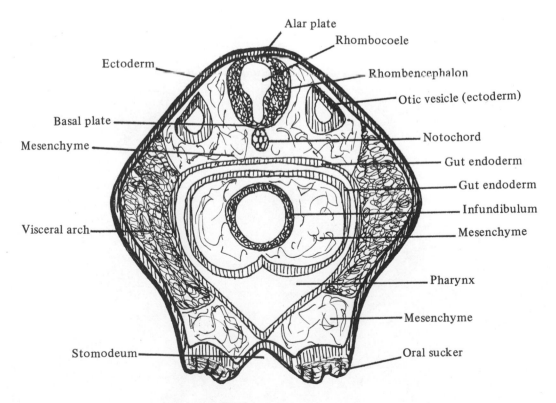

Alar plate
Rhombocoele
Rhombencephalon
Otic vesicle (ectoderm)
Notochord
Gut endoderm
Gut endoderm
Infundibulum
Mesenchyme
Pharynx
Mesenchyme
Oral sucker

Ectoderm
Basal plate
Mesenchyme
Visceral arch
Stomodeum

Section 4

Frog development   4 mm. embryo

x 100, x. sec.

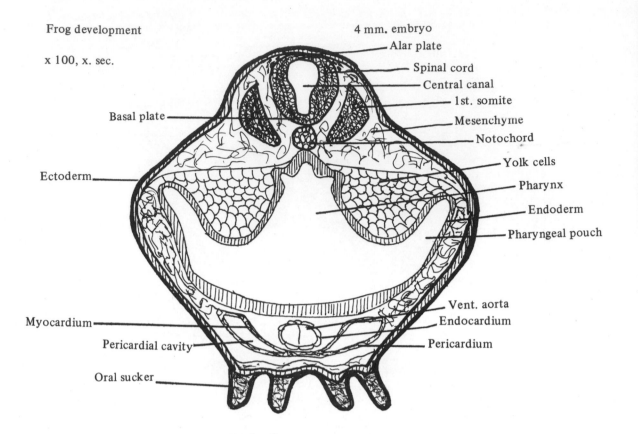

Alar plate
Spinal cord
Central canal
1st. somite
Basal plate
Mesenchyme
Notochord
Ectoderm
Yolk cells
Pharynx
Endoderm
Pharyngeal pouch

Myocardium
Vent. aorta
Endocardium
Pericardial cavity
Pericardium
Oral sucker

Section 5

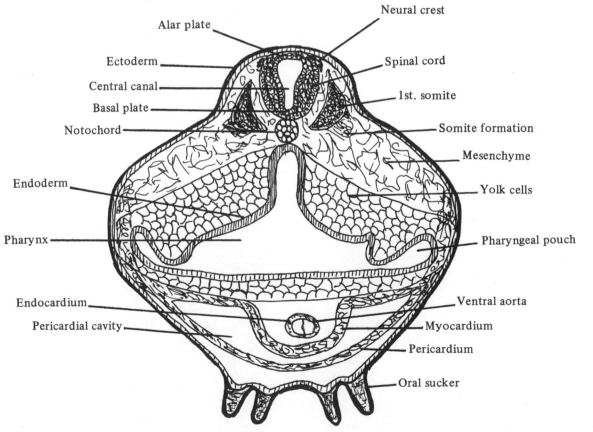

Neural crest
Alar plate
Ectoderm
Spinal cord
Central canal
Basal plate
1st. somite
Notochord
Somite formation
Mesenchyme
Endoderm
Yolk cells
Pharynx
Pharyngeal pouch
Endocardium
Ventral aorta
Pericardial cavity
Myocardium
Pericardium
Oral sucker

Section 6

Frog development

4 mm. embryo

x 100. x. sec.

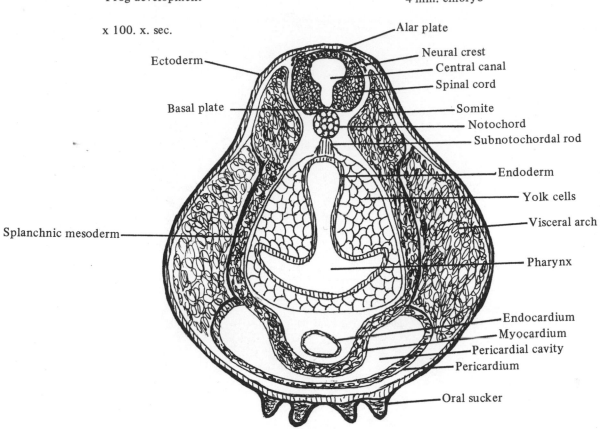

Alar plate
Neural crest
Central canal
Spinal cord
Ectoderm
Somite
Basal plate
Notochord
Subnotochordal rod
Endoderm
Yolk cells
Visceral arch
Splanchnic mesoderm
Pharynx
Endocardium
Myocardium
Pericardial cavity
Pericardium
Oral sucker

Section 7

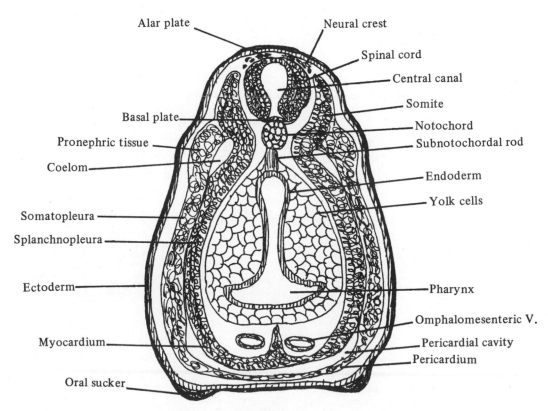

Alar plate
Neural crest
Spinal cord
Central canal
Basal plate
Somite
Pronephric tissue
Notochord
Coelom
Subnotochordal rod
Endoderm
Somatopleura
Yolk cells
Splanchnopleura
Ectoderm
Pharynx
Omphalomesenteric V.
Myocardium
Pericardial cavity
Pericardium
Oral sucker

Section 8

Frog development

4 mm. embryo

x 100, x. sec.

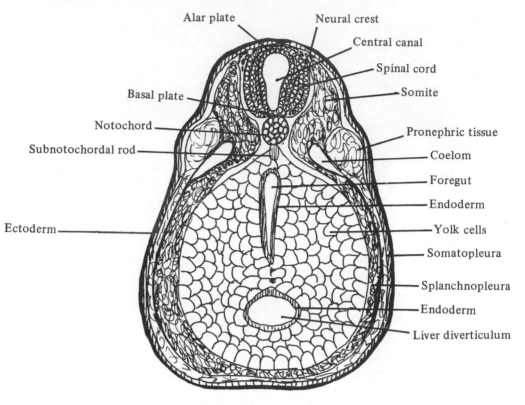

Alar plate   Neural crest

Central canal

Spinal cord

Basal plate   Somite

Notochord   Pronephric tissue

Subnotochordal rod   Coelom

Foregut

Endoderm

Ectoderm   Yolk cells

Somatopleura

Splanchnopleura

Endoderm

Liver diverticulum

Section 9

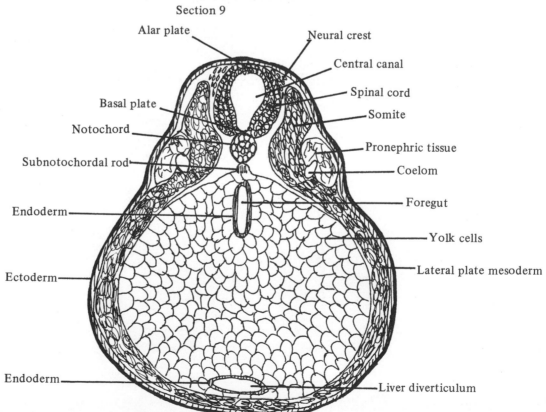

Alar plate   Neural crest

Central canal

Spinal cord

Basal plate   Somite

Notochord   Pronephric tissue

Subnotochordal rod   Coelom

Foregut

Endoderm

Yolk cells

Ectoderm   Lateral plate mesoderm

Endoderm   Liver diverticulum

Section 10

Frog development

x 100, x. sec.

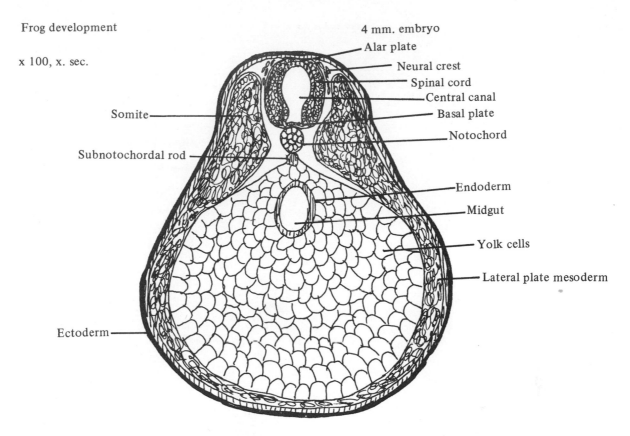

4 mm. embryo
Alar plate
Neural crest
Spinal cord
Central canal
Basal plate
Notochord

Somite

Subnotochordal rod

Endoderm
Midgut
Yolk cells
Lateral plate mesoderm

Ectoderm

Section 11

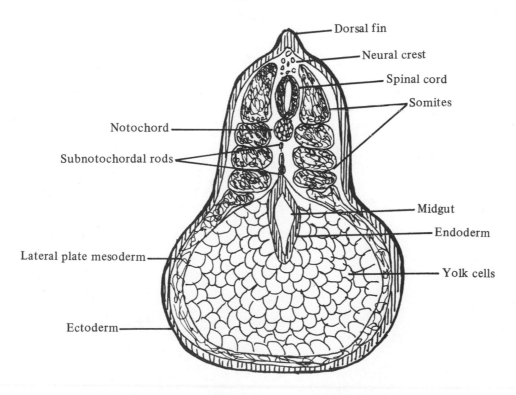

Dorsal fin
Neural crest
Spinal cord
Somites
Notochord
Subnotochordal rods
Midgut
Endoderm
Lateral plate mesoderm
Yolk cells
Ectoderm

Section 12

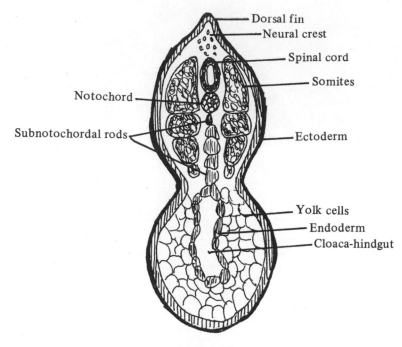

Dorsal fin
Neural crest
Spinal cord
Somites
Notochord
Subnotochordal rods
Ectoderm
Yolk cells
Endoderm
Cloaca-hindgut

Section 13

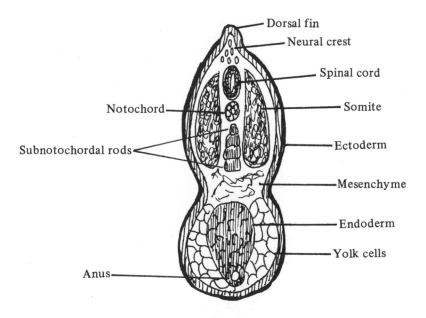

Dorsal fin
Neural crest
Spinal cord
Notochord
Somite
Subnotochordal rods
Ectoderm
Mesenchyme
Endoderm
Yolk cells
Anus

Section 14

90

Frog development   4 mm. embryo

x 100, x. sec.

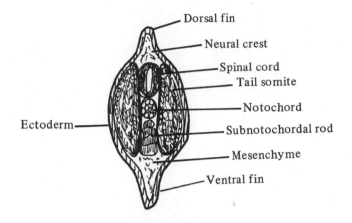

Dorsal fin

Neural crest

Spinal cord

Tail somite

Notochord

Subnotochordal rod

Ectoderm

Mesenchyme

Ventral fin

Section 15

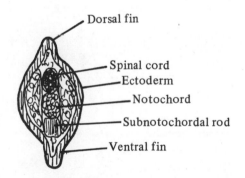

Dorsal fin

Spinal cord

Ectoderm

Notochord

Subnotochordal rod

Ventral fin

Section 16

Frog development   **7** mm. embryo

x 100, median sag. sec.

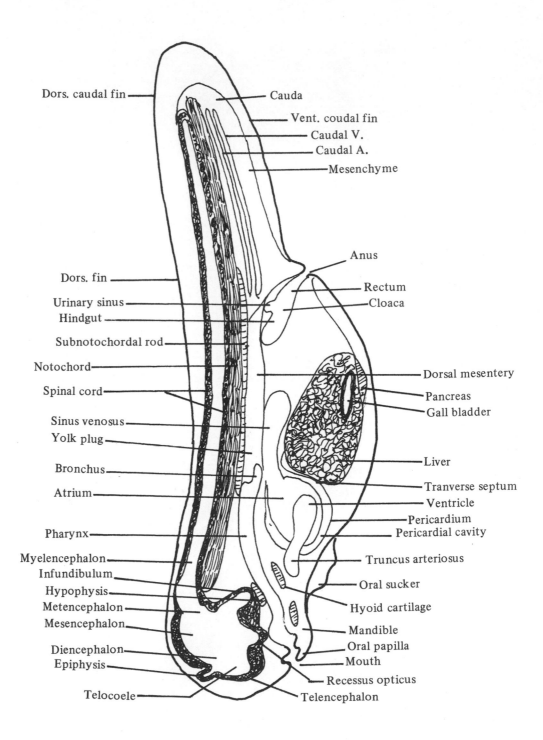

Dors. caudal fin ——————————— Cauda

—————— Vent. coudal fin
—————— Caudal V.
—————— Caudal A.
—————— Mesenchyme

—————— Anus

Dors. fin ——————————— Rectum
Urinary sinus ——————— Cloaca
Hindgut ———————————

Subnotochordal rod ———

Notochord ————————— Dorsal mesentery
Spinal cord ————————— Pancreas
————— Gall bladder

Sinus venosus ————————
Yolk plug ———————————

Bronchus ——————————— Liver

Atrium ——————————— Tranverse septum
————— Ventricle
————— Pericardium
Pharynx ——————————— Pericardial cavity

Myelencephalon ——————— Truncus arteriosus
Infundibulum ————————
Hypophysis ————————— Oral sucker
Metencephalon —————————
Mesencephalon ———————— Hyoid cartilage

Diencephalon ————————— Mandible
Epiphysis ————————— Oral papilla
————— Mouth
————— Recessus opticus
Telocoele ————————— Telencephalon

Frog development   7 mm. embryo

x 100, median sag. sec.

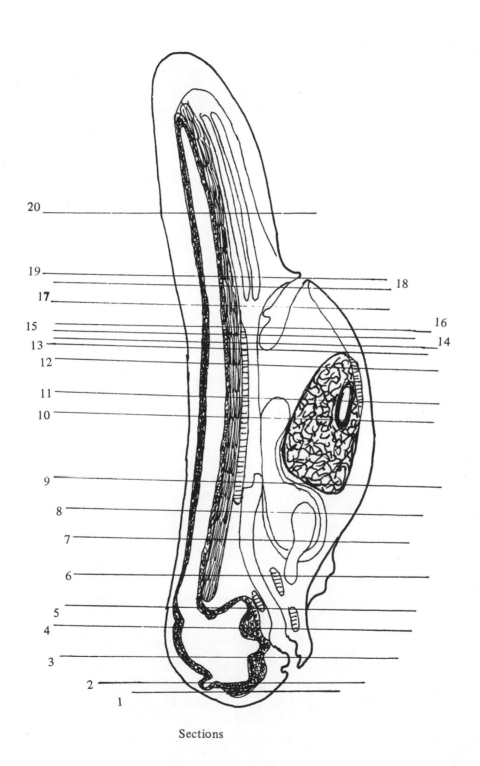

Sections

Frog development
7 mm. embryo

x 100, x. sec.

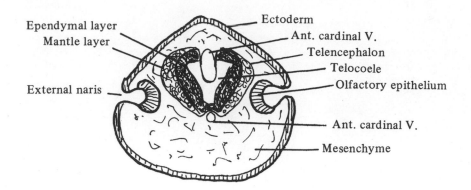

Ependymal layer — Ectoderm
Mantle layer — Ant. cardinal V.
— Telencephalon
— Telocoele
External naris — Olfactory epithelium

— Ant. cardinal V.

— Mesenchyme

Section 1

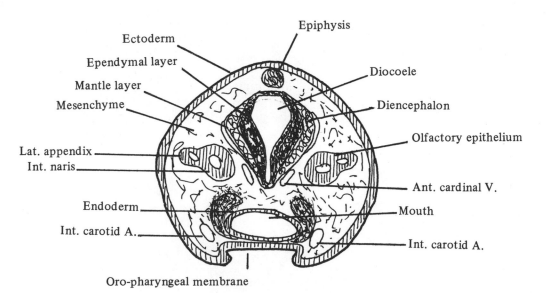

Ectoderm — Epiphysis

Ependymal layer

Mantle layer — Diocoele

Mesenchyme — Diencephalon

— Olfactory epithelium

Lat. appendix —
Int. naris —

— Ant. cardinal V.

Endoderm — Mouth

Int. carotid A. — Int. carotid A.

Oro-pharyngeal membrane

Section 2

Frog development   7 mm. embryo

x 100, x. sec.

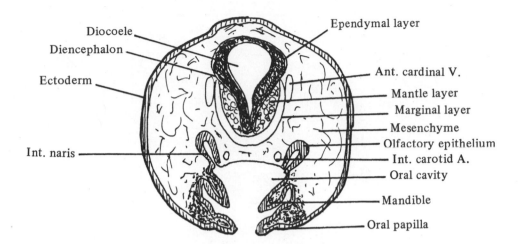

Diocoele
Diencephalon
Ectoderm
Int. naris

Ependymal layer
Ant. cardinal V.
Mantle layer
Marginal layer
Mesenchyme
Olfactory epithelium
Int. carotid A.
Oral cavity
Mandible
Oral papilla

Section 3

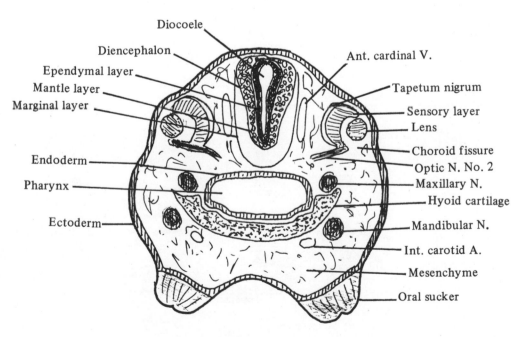

Diocoele
Diencephalon
Ependymal layer
Mantle layer
Marginal layer
Endoderm
Pharynx
Ectoderm

Ant. cardinal V.
Tapetum nigrum
Sensory layer
Lens
Choroid fissure
Optic N. No. 2
Maxillary N.
Hyoid cartilage
Mandibular N.
Int. carotid A.
Mesenchyme
Oral sucker

Section 4

Frog development   **7** mm. embryo

x 100, x. sec.

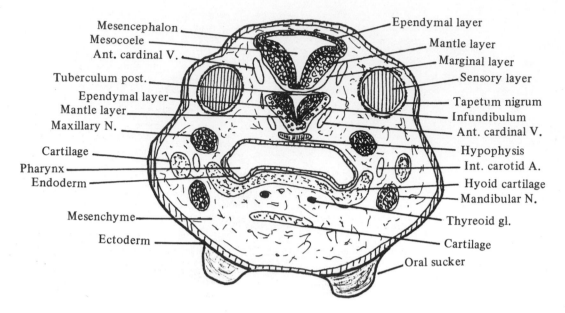

Mesencephalon
Mesocoele
Ant. cardinal V.
Tuberculum post.
Ependymal layer
Mantle layer
Maxillary N.
Cartilage
Pharynx
Endoderm
Mesenchyme
Ectoderm

Ependymal layer
Mantle layer
Marginal layer
Sensory layer
Tapetum nigrum
Infundibulum
Ant. cardinal V.
Hypophysis
Int. carotid A.
Hyoid cartilage
Mandibular N.
Thyreoid gl.
Cartilage
Oral sucker

Section 5

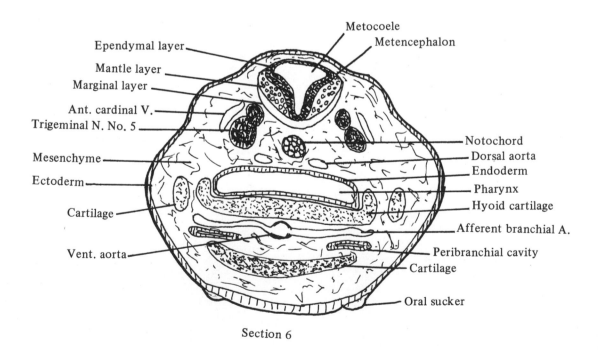

Ependymal layer
Mantle layer
Marginal layer
Ant. cardinal V.
Trigeminal N. No. 5
Mesenchyme
Ectoderm
Cartilage
Vent. aorta

Metocoele
Metencephalon
Notochord
Dorsal aorta
Endoderm
Pharynx
Hyoid cartilage
Afferent branchial A.
Peribranchial cavity
Cartilage
Oral sucker

Section 6

Frog development

7 mm. embryo

x 100, x. sec.

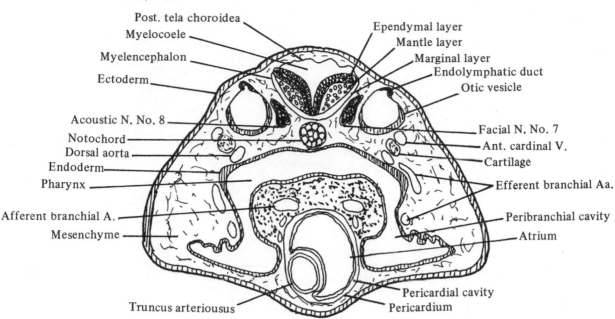

Post. tela choroidea
Myelocoele
Myelencephalon
Ectoderm

Ependymal layer
Mantle layer
Marginal layer
Endolymphatic duct
Otic vesicle

Acoustic N. No. 8
Notochord
Dorsal aorta
Endoderm
Pharynx

Facial N. No. 7
Ant. cardinal V.
Cartilage

Efferent branchial Aa.

Afferent branchial A.
Mesenchyme

Peribranchial cavity
Atrium

Truncus arteriousus

Pericardial cavity
Pericardium

Section 7

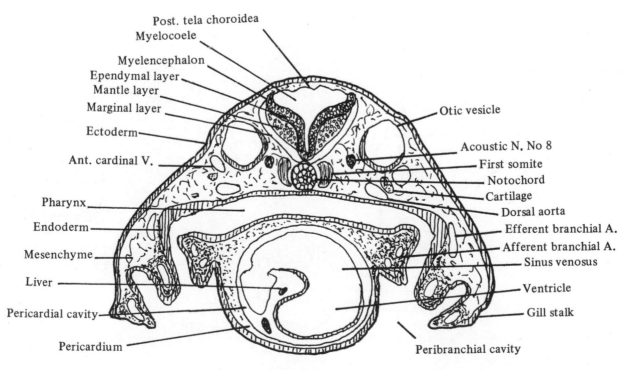

Post. tela choroidea
Myelocoele

Myelencephalon
Ependymal layer
Mantle layer
Marginal layer
Ectoderm
Ant. cardinal V.

Otic vesicle

Acoustic N, No 8
First somite
Notochord
Cartilage
Dorsal aorta
Efferent branchial A.
Afferent branchial A.
Sinus venosus

Pharynx
Endoderm

Mesenchyme

Liver

Ventricle

Pericardial cavity

Gill stalk

Pericardium

Peribranchial cavity

Section 8

x 100, x. sec.

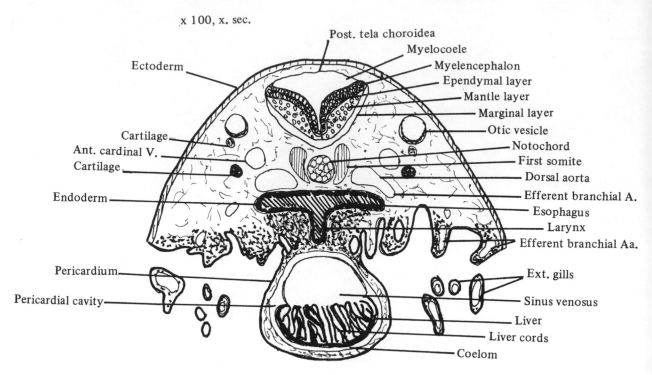

Post. tela choroidea
Myelocoele
Myelencephalon
Ependymal layer
Mantle layer
Marginal layer
Otic vesicle
Notochord
First somite
Dorsal aorta
Efferent branchial A.
Esophagus
Larynx
Efferent branchial Aa.
Ext. gills
Sinus venosus
Liver
Liver cords
Coelom

Ectoderm
Cartilage
Ant. cardinal V.
Cartilage
Endoderm
Pericardium
Pericardial cavity

Section 9

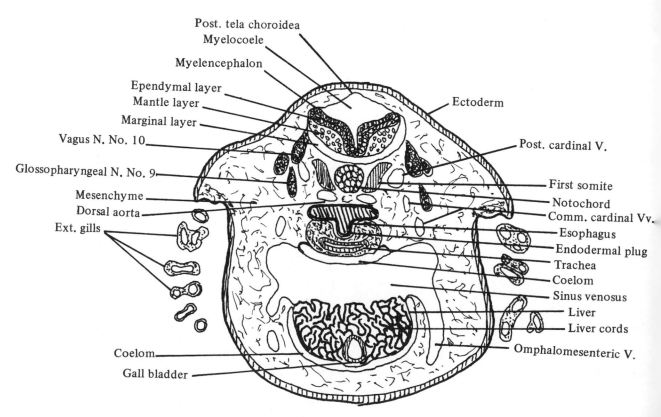

Post. tela choroidea
Myelocoele
Myelencephalon
Ependymal layer
Mantle layer
Marginal layer
Vagus N. No. 10
Glossopharyngeal N. No. 9
Mesenchyme
Dorsal aorta
Ext. gills
Coelom
Gall bladder

Ectoderm
Post. cardinal V.
First somite
Notochord
Comm. cardinal Vv.
Esophagus
Endodermal plug
Trachea
Coelom
Sinus venosus
Liver
Liver cords
Omphalomesenteric V.

Section 10

Frog development

7 mm. embryo

x 100, x. sec.

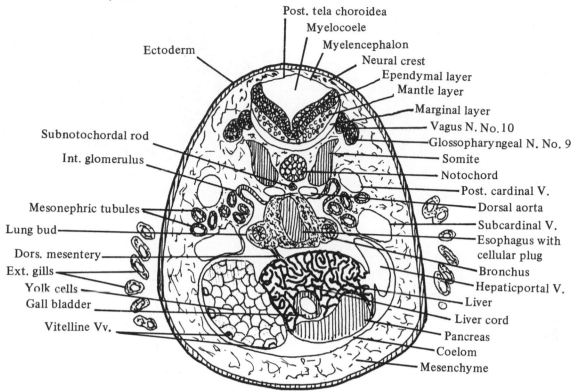

Post. tela choroidea
Myelocoele
Myelencephalon
Neural crest
Ependymal layer
Mantle layer

Ectoderm

Marginal layer
Vagus N. No. 10
Glossopharyngeal N. No. 9
Somite
Notochord
Post. cardinal V.
Dorsal aorta
Subcardinal V.
Esophagus with cellular plug
Bronchus
Hepaticportal V.
Liver
Liver cord
Pancreas
Coelom
Mesenchyme

Subnotochordal rod
Int. glomerulus

Mesonephric tubules
Lung bud
Dors. mesentery
Ext. gills
Yolk cells
Gall bladder
Vitelline Vv.

Section 11

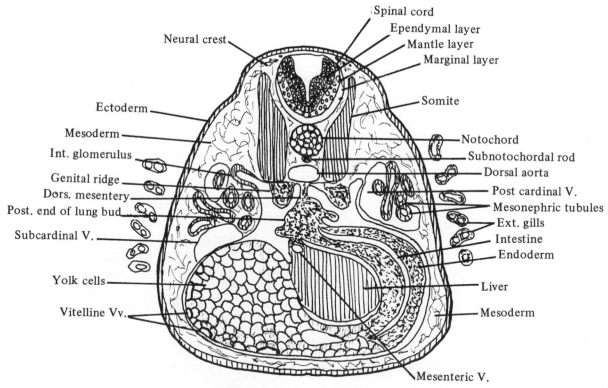

Spinal cord
Ependymal layer
Mantle layer
Marginal layer

Neural crest

Ectoderm
Mesoderm
Int. glomerulus
Genital ridge
Dors. mesentery
Post. end of lung bud
Subcardinal V.
Yolk cells
Vitelline Vv.

Somite
Notochord
Subnotochordal rod
Dorsal aorta
Post cardinal V.
Mesonephric tubules
Ext. gills
Intestine
Endoderm
Liver
Mesoderm

Mesenteric V.

Section 12

Frog development

x 100, x. sec.

7 mm. embryo

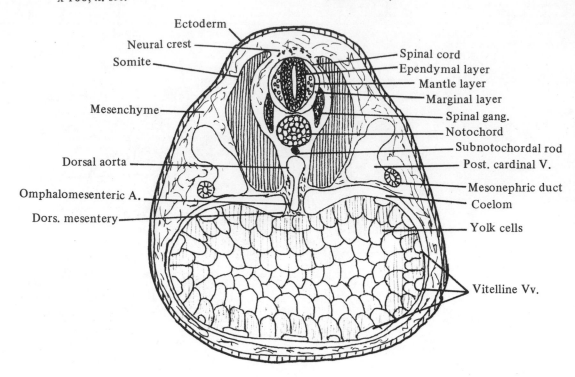

Ectoderm

Neural crest

Somite

Mesenchyme

Dorsal aorta

Omphalomesenteric A.

Dors. mesentery

Spinal cord

Ependymal layer

Mantle layer

Marginal layer

Spinal gang.

Notochord

Subnotochordal rod

Post. cardinal V.

Mesonephric duct

Coelom

Yolk cells

Vitelline Vv.

Section 13

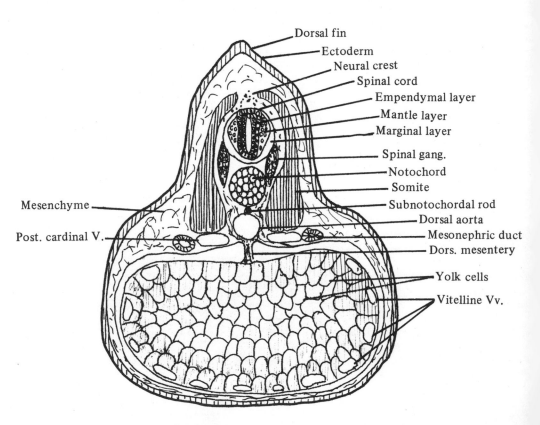

Dorsal fin

Ectoderm

Neural crest

Spinal cord

Empendymal layer

Mantle layer

Marginal layer

Spinal gang.

Notochord

Somite

Subnotochordal rod

Dorsal aorta

Mesonephric duct

Dors. mesentery

Yolk cells

Vitelline Vv.

Mesenchyme

Post. cardinal V.

Section 14

Frog development
x 100, x. sec.

7 mm. embryo

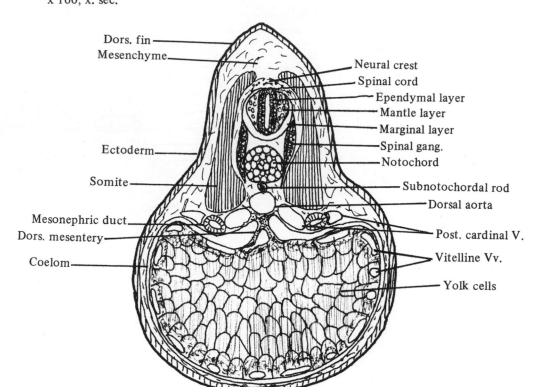

Dors. fin
Mesenchyme

Neural crest
Spinal cord
Ependymal layer
Mantle layer
Marginal layer
Spinal gang.
Notochord

Ectoderm

Somite

Subnotochordal rod
Dorsal aorta

Mesonephric duct
Dors. mesentery

Post. cardinal V.

Coelom

Vitelline Vv.

Yolk cells

Section 15

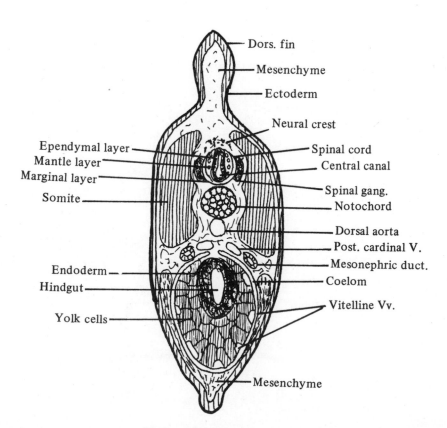

Dors. fin

Mesenchyme

Ectoderm

Neural crest

Ependymal layer
Mantle layer
Marginal layer

Spinal cord
Central canal

Somite

Spinal gang.
Notochord

Dorsal aorta
Post. cardinal V.
Mesonephric duct.

Endoderm
Hindgut

Coelom

Vitelline Vv.

Yolk cells

Mesenchyme

Section 16

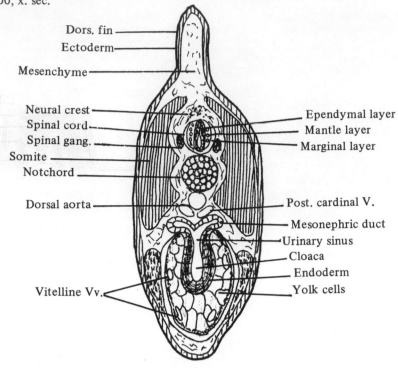

Dors. fin

Ectoderm

Mesenchyme

Neural crest — Ependymal layer

Spinal cord — Mantle layer

Spinal gang. — Marginal layer

Somite

Notchord

Dorsal aorta — Post. cardinal V.

Mesonephric duct

Urinary sinus

Cloaca

Endoderm

Vitelline Vv. — Yolk cells

Section 17

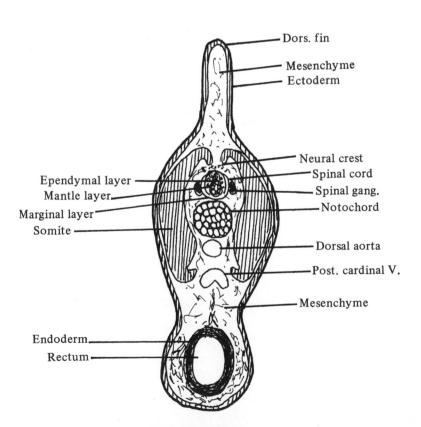

Dors. fin

Mesenchyme

Ectoderm

Neural crest

Ependymal layer — Spinal cord

Mantle layer — Spinal gang.

Marginal layer — Notochord

Somite

Dorsal aorta

Post. cardinal V.

Mesenchyme

Endoderm

Rectum

Section 18

Frog development

7 mm. embryo

x 100, x. sec.

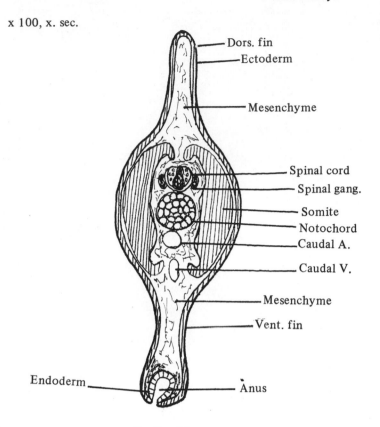

Dors. fin
Ectoderm

Mesenchyme

Spinal cord
Spinal gang.
Somite
Notochord
Caudal A.

Caudal V.

Mesenchyme

Vent. fin

Endoderm

Anus

Section 19

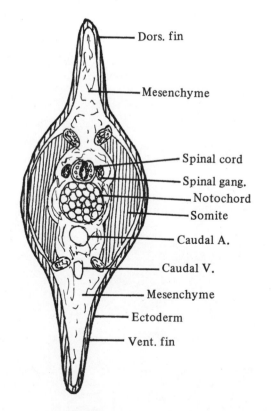

Dors. fin

Mesenchyme

Spinal cord
Spinal gang.
Notochord
Somite
Caudal A.

Caudal V.

Mesenchyme
Ectoderm
Vent. fin

Section 20

Part II

Development of the Chick Embryo I

Chick Testis
Chick Ovary
Chick Early Cleavage

Development of the chick embryo I
Chick testis
x 100 x. sec.

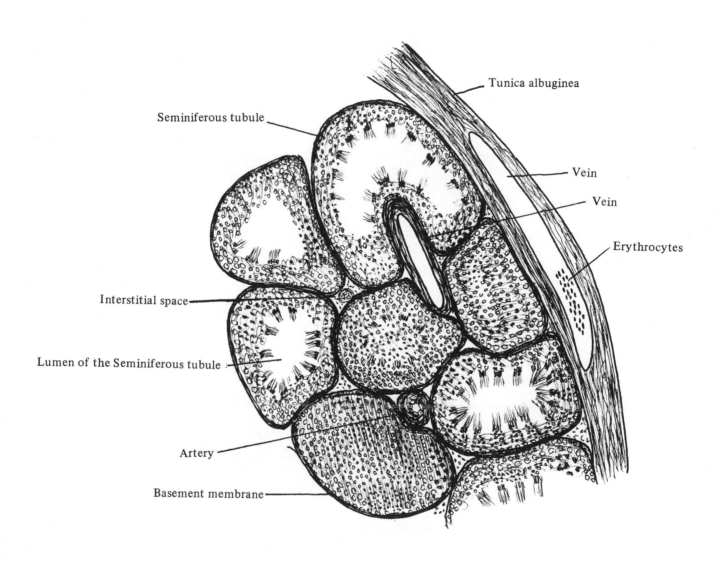

Tunica albuginea

Seminiferous tubule

Vein

Vein

Erythrocytes

Interstitial space

Lumen of the Seminiferous tubule

Artery

Basement membrane

Development of the chick embryo I
   Chick testis
   x 1.000 x. sec

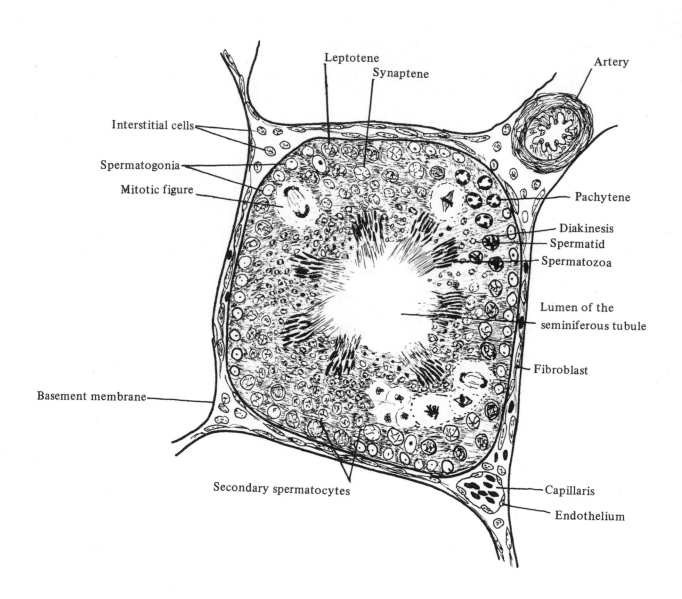

Leptotene
Synaptene
Artery
Interstitial cells
Spermatogonia
Mitotic figure
Pachytene
Diakinesis
Spermatid
Spermatozoa
Lumen of the
seminiferous tubule
Fibroblast
Basement membrane
Secondary spermatocytes
Capillaris
Endothelium

Development of the chick embryo I
Chick ovary
x 100 x. sec.

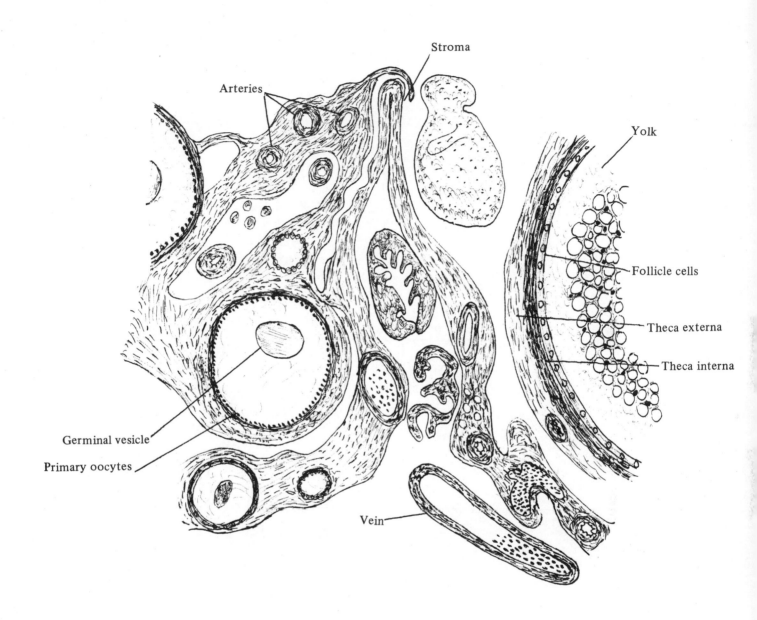

Arteries

Stroma

Yolk

Follicle cells

Theca externa

Theca interna

Germinal vesicle

Primary oocytes

Vein

Development of the chick embryo I
Chick ovary, younger oocytes
x 1.000 x. sec.

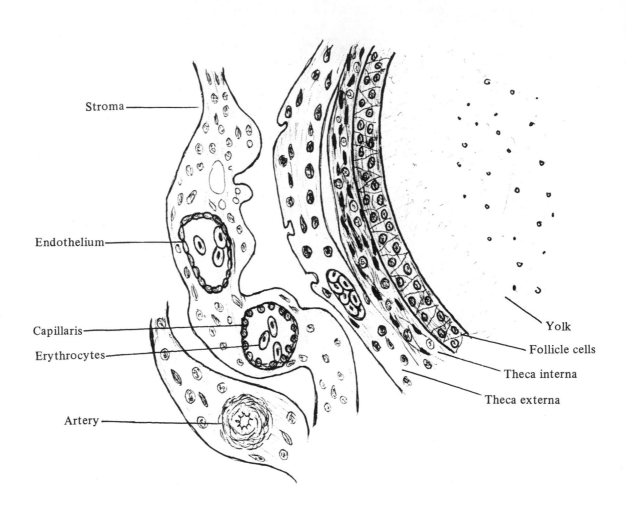

Stroma

Endothelium

Capillaris

Erythrocytes

Artery

Yolk

Follicle cells

Theca interna

Theca externa

Development of the chick embryo I
Chick ovary, older oocyte
x 1.000 x. sec.

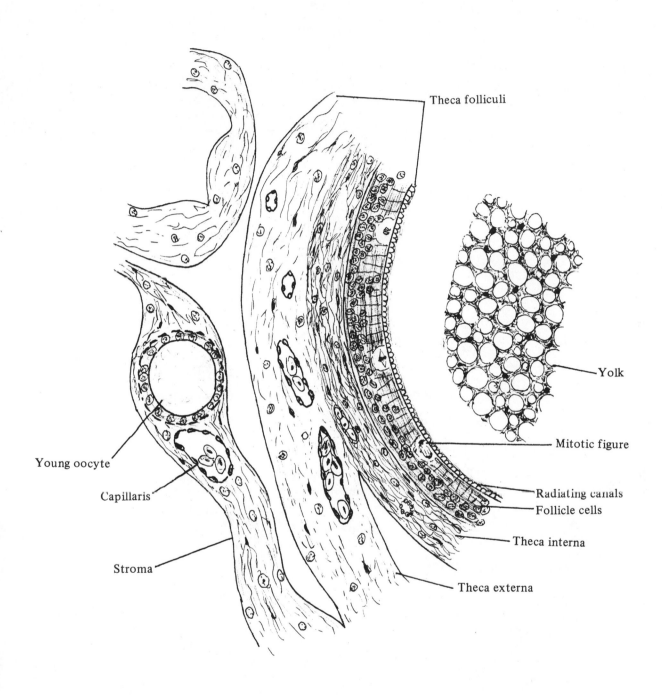

Theca folliculi

Yolk

Mitotic figure

Young oocyte

Capillaris

Radiating canals

Follicle cells

Theca interna

Stroma

Theca externa

Development of the chick embryo I
Early cleavage
x 400 Median sec.

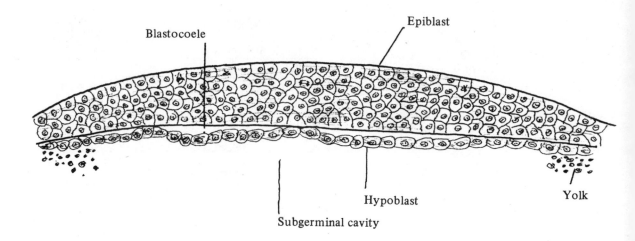

Blastocoele

Epiblast

Hypoblast

Yolk

Subgerminal cavity

Development of the Chick Embryo II
18-Hour Chick Embryo

Development of the chick embryo II
x 100 w.m.

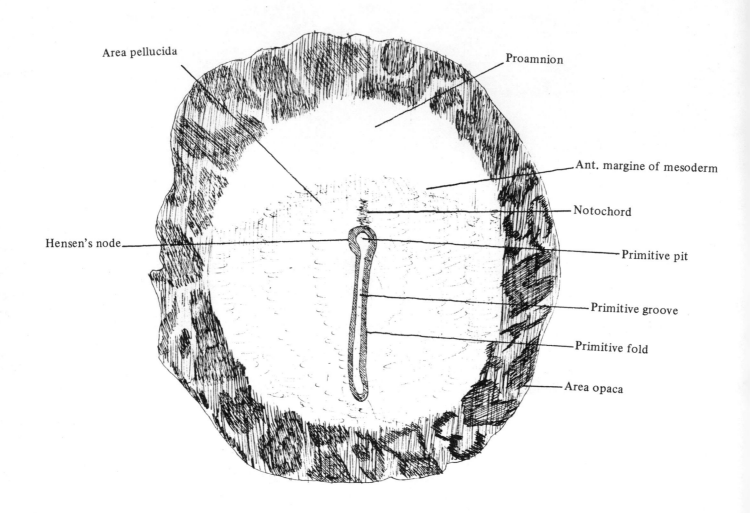

Area pellucida

Proamnion

Ant. margine of mesoderm

Notochord

Hensen's node

Primitive pit

Primitive groove

Primitive fold

Area opaca

114

Development of the chick embryo II
x 100 w.m.

Sections

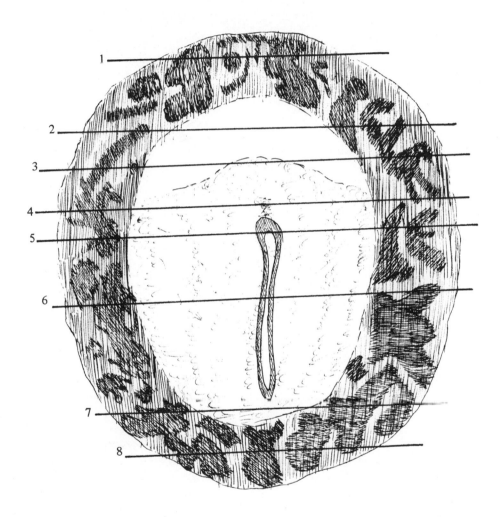

1

2

3

4

5

6

7

8

Development of the chick embryo II
x 100 x. sec.

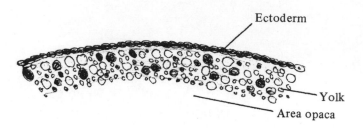

Section 1.

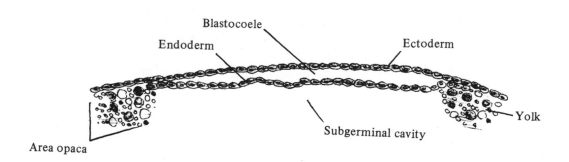

Section 2.

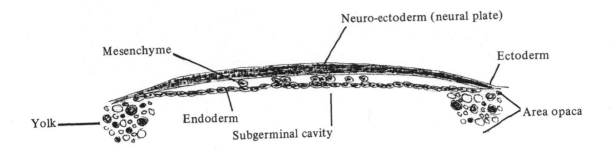

Neuro-ectoderm (neural plate)

Mesenchyme

Ectoderm

Yolk

Endoderm

Area opaca

Subgerminal cavity

Section 3.

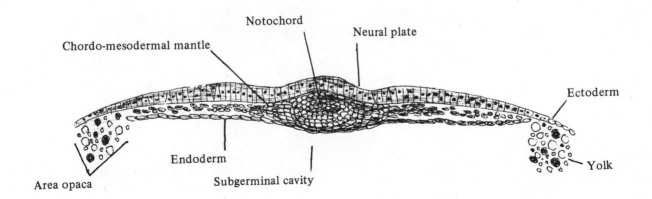

Notochord

Neural plate

Chordo-mesodermal mantle

Ectoderm

Endoderm

Yolk

Area opaca

Subgerminal cavity

Section 4.

Development of the chick embryo II
x 100 x. sec.

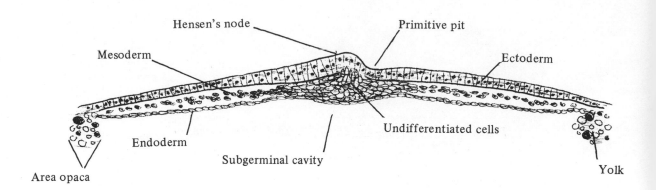

Section 5.

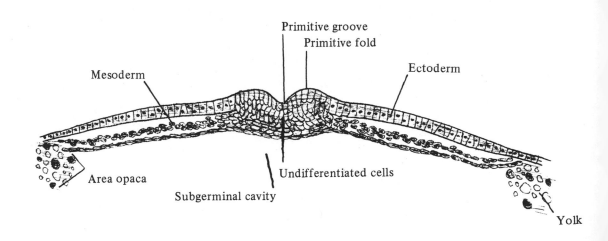

Section 6.

Development of the chick embryo II
x 100 x. sec.

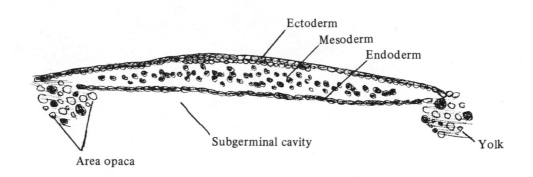

Section 7.

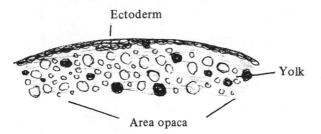

Section 8.

Development of the chick embryo II
x 1.000 x. sec.

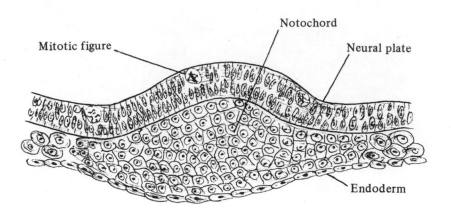

Mitotic figure

Notochord

Neural plate

Endoderm

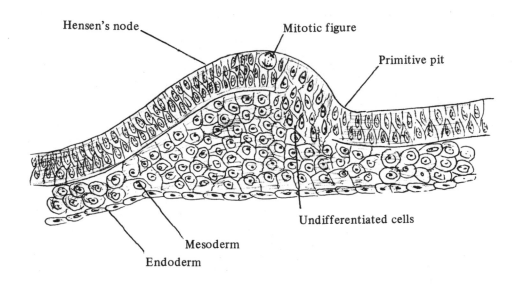

Hensen's node

Mitotic figure

Primitive pit

Undifferentiated cells

Mesoderm

Endoderm

Development of the Chick Embryo III
24-Hour Chick Embryo

Development of the chick embryo III
x 100 w.m.

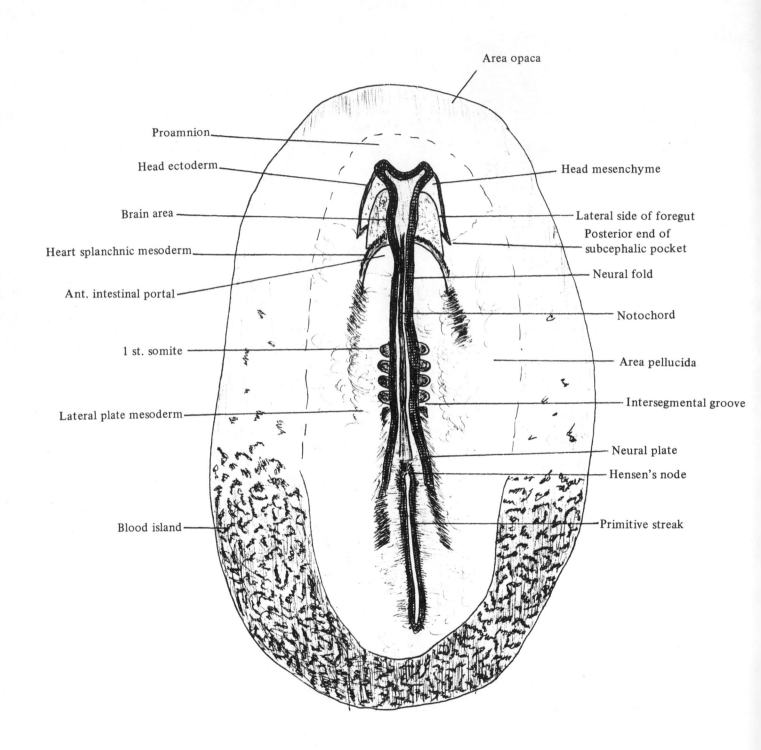

Area opaca

Proamnion

Head ectoderm

Head mesenchyme

Brain area

Lateral side of foregut

Heart splanchnic mesoderm

Posterior end of
subcephalic pocket

Ant. intestinal portal

Neural fold

Notochord

1 st. somite

Area pellucida

Intersegmental groove

Lateral plate mesoderm

Neural plate

Hensen's node

Blood island

Primitive streak

122

Development of the chick embryo III
x 100 w.m.

A

Sections·

1
2
3
4
5

6

7

8

9

10

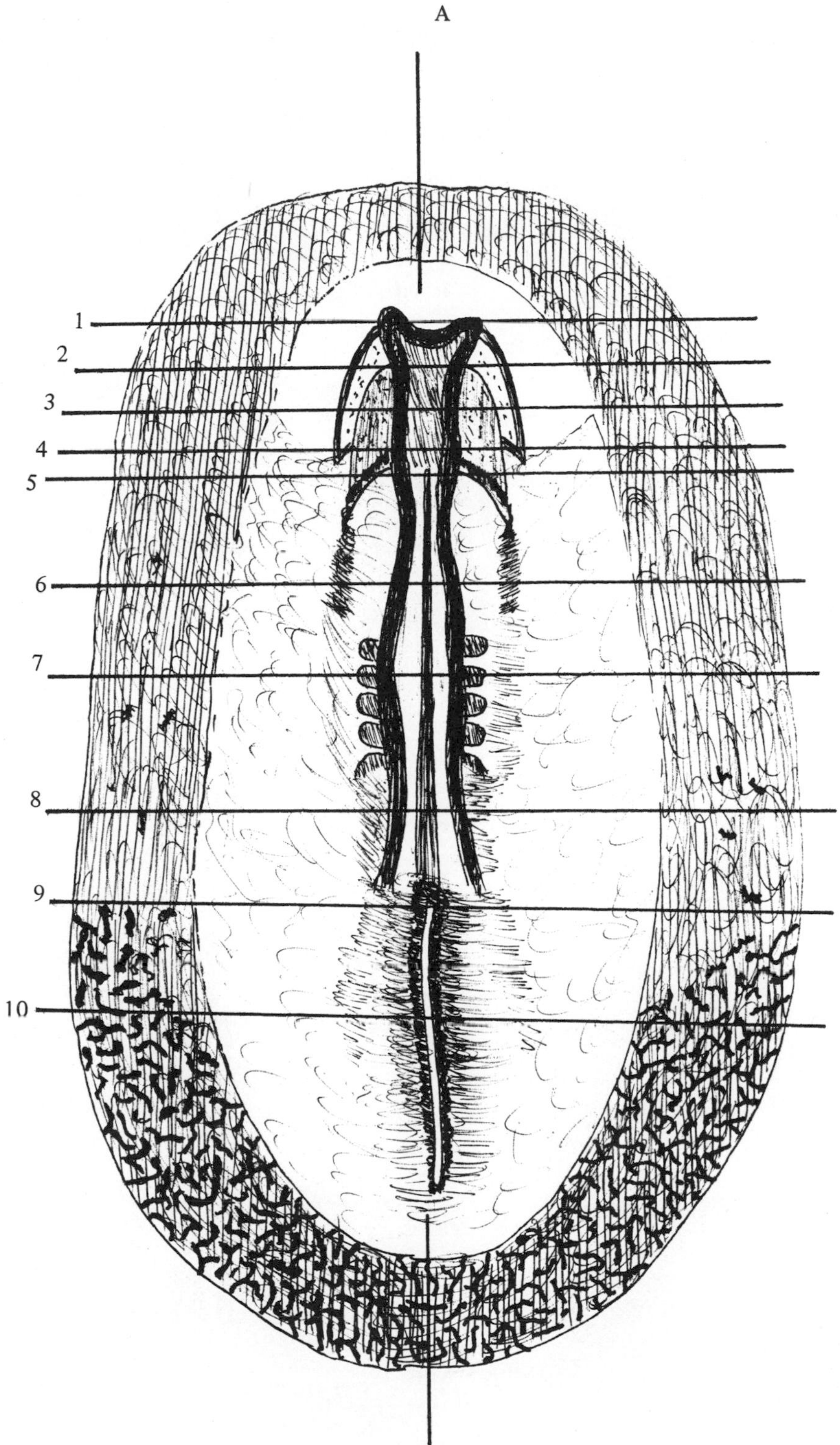

Development of the chick embryo III
x 100 mid. sag. sec.

Section A.

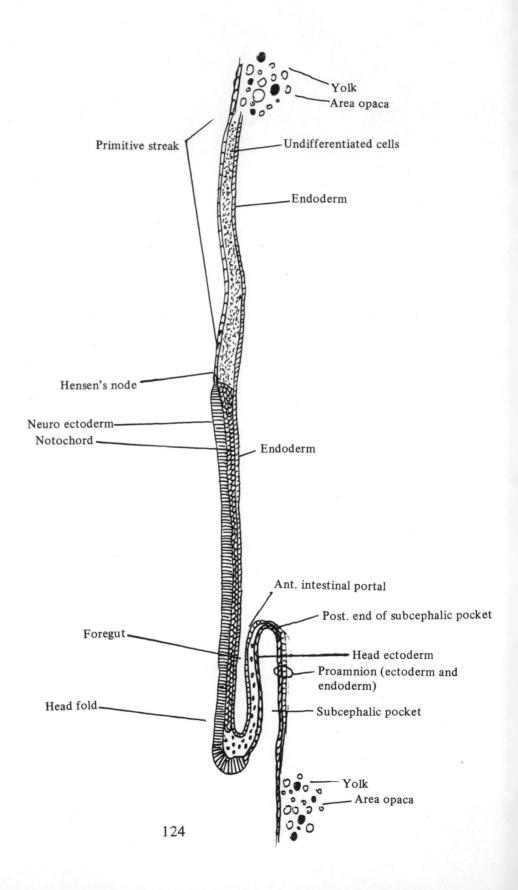

Yolk

Area opaca

Primitive streak

Undifferentiated cells

Endoderm

Hensen's node

Neuro ectoderm

Notochord

Endoderm

Ant. intestinal portal

Post. end of subcephalic pocket

Foregut

Head ectoderm

Proamnion (ectoderm and endoderm)

Head fold

Subcephalic pocket

Yolk

Area opaca

Development of the chick embryo III
x 100 x. sec.

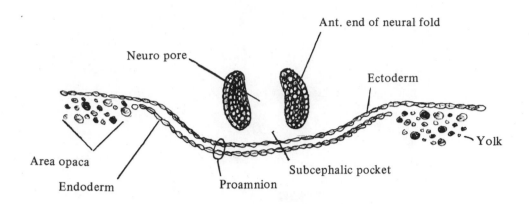

Ant. end of neural fold

Neuro pore

Ectoderm

Area opaca

Endoderm

Proamnion

Subcephalic pocket

Yolk

Section 1.

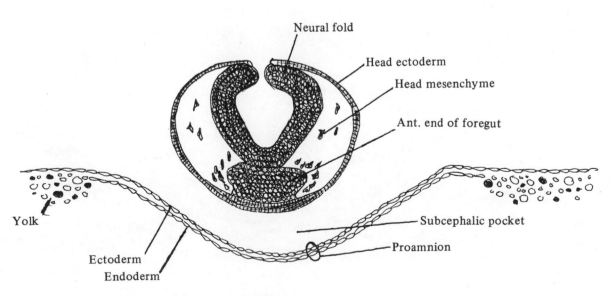

Neural fold

Head ectoderm

Head mesenchyme

Ant. end of foregut

Yolk

Ectoderm

Endoderm

Subcephalic pocket

Proamnion

Section 2.

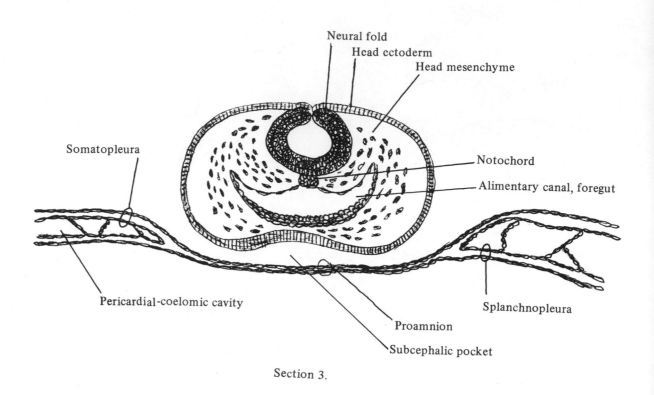

Section 3.

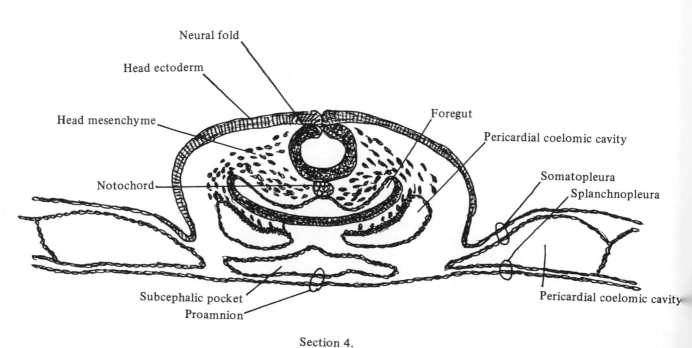

Section 4.

Development of the chick embryo III
x 100 x. sec.

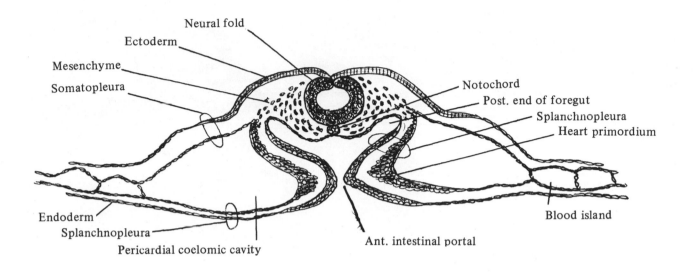

Neural fold

Ectoderm

Mesenchyme

Somatopleura

Notochord

Post. end of foregut

Splanchnopleura

Heart primordium

Endoderm

Splanchnopleura

Pericardial coelomic cavity

Ant. intestinal portal

Blood island

Section 5.

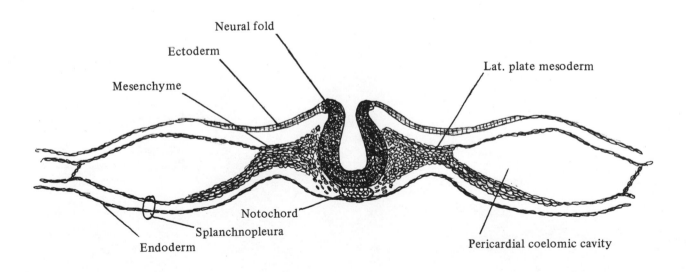

Neural fold

Ectoderm

Mesenchyme

Lat. plate mesoderm

Notochord

Splanchnopleura

Endoderm

Pericardial coelomic cavity

Section 6.

Development of the chick embryo III
x 100 x. sec.

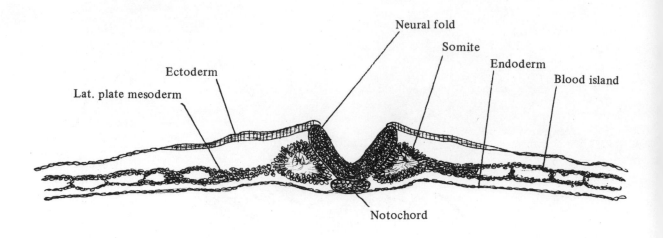

Lat. plate mesoderm   Ectoderm   Neural fold   Somite   Endoderm   Blood island

Notochord

Section 7

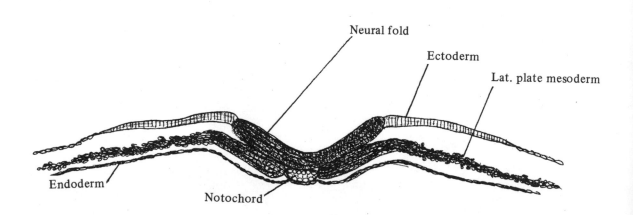

Neural fold   Ectoderm   Lat. plate mesoderm

Endoderm   Notochord

Section 8

Development of the chick embryo III
x 100 x. sec.

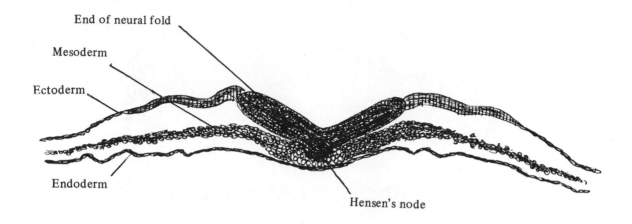

End of neural fold

Mesoderm

Ectoderm

Endoderm

Hensen's node

Section 9

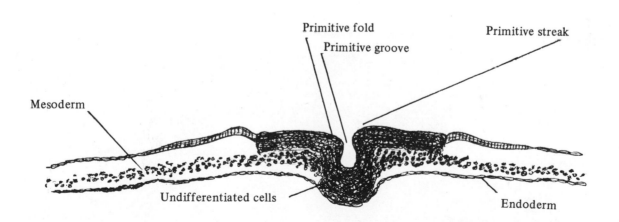

Primitive fold

Primitive groove

Primitive streak

Mesoderm

Undifferentiated cells

Endoderm

Section 10

Development of the Chick Embryo IV

33-Hour Chick Embryo

Development of the chick embryo IV
x 40 w.m.

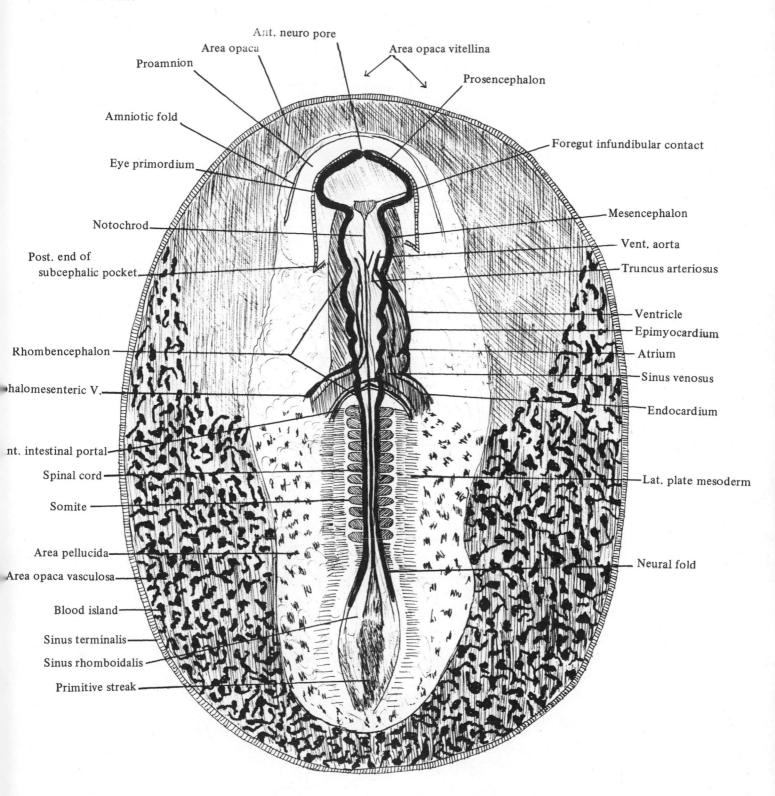

Ant. neuro pore

Area opaca

Area opaca vitellina

Proamnion

Prosencephalon

Amniotic fold

Foregut infundibular contact

Eye primordium

Mesencephalon

Notochrod

Vent. aorta

Post. end of
subcephalic pocket

Truncus arteriosus

Ventricle

Epimyocardium

Rhombencephalon

Atrium

Sinus venosus

halomesenteric V.

Endocardium

nt. intestinal portal

Lat. plate mesoderm

Spinal cord

Somite

Area pellucida

Neural fold

Area opaca vasculosa

Blood island

Sinus terminalis

Sinus rhomboidalis

Primitive streak

Development of the chick embryo IV
x 40 w.m.

Sections

A

1
2
3
4
5
6
7
8
9
10
11
12

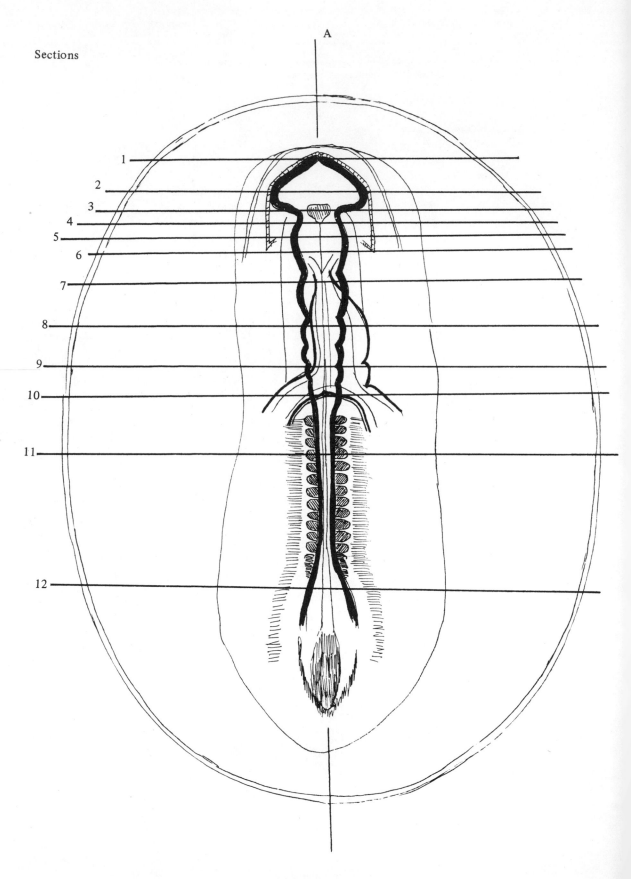

Development of the chick embryo IV
x 100 mid.sag.sec.

Section A.

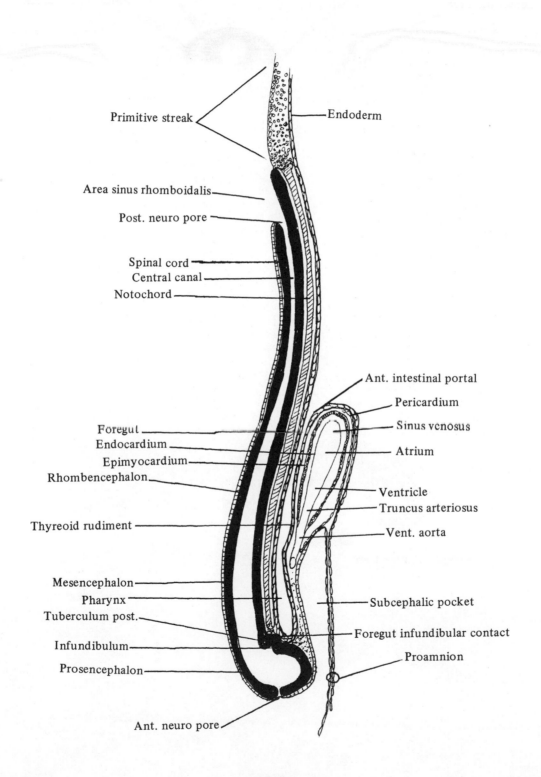

Primitive streak — Endoderm

Area sinus rhomboidalis

Post. neuro pore

Spinal cord
Central canal
Notochord

Ant. intestinal portal
Pericardium
Foregut — Sinus venosus
Endocardium
Epimyocardium — Atrium
Rhombencephalon

Thyreoid rudiment — Ventricle
Truncus arteriosus
Vent. aorta

Mesencephalon
Pharynx
Tuberculum post. — Subcephalic pocket

Foregut infundibular contact
Infundibulum
Prosencephalon — Proamnion

Ant. neuro pore

Development of the chick embryo IV
x 100 x. sec.

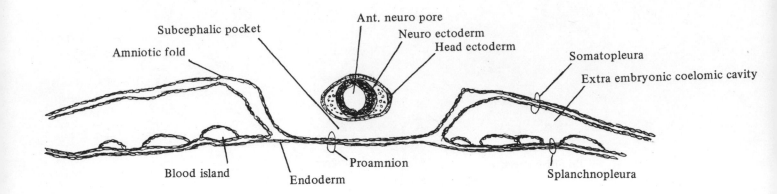

Subcephalic pocket

Amniotic fold

Ant. neuro pore

Neuro ectoderm

Head ectoderm

Somatopleura

Extra embryonic coelomic cavity

Blood island

Endoderm

Proamnion

Splanchnopleura

Section 1

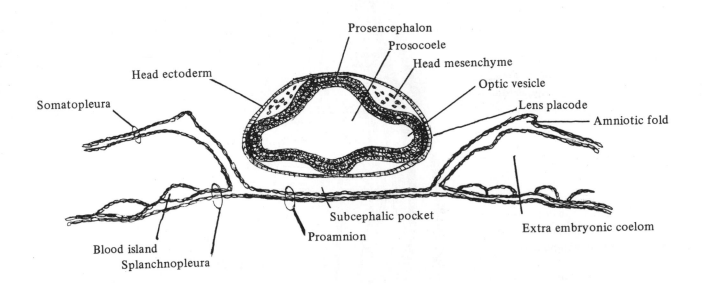

Prosencephalon

Prosocoele

Head mesenchyme

Head ectoderm

Optic vesicle

Lens placode

Somatopleura

Amniotic fold

Subcephalic pocket

Proamnion

Extra embryonic coelom

Blood island

Splanchnopleura

Section 2

Development of the chick embryo IV
x 100 x. sec.

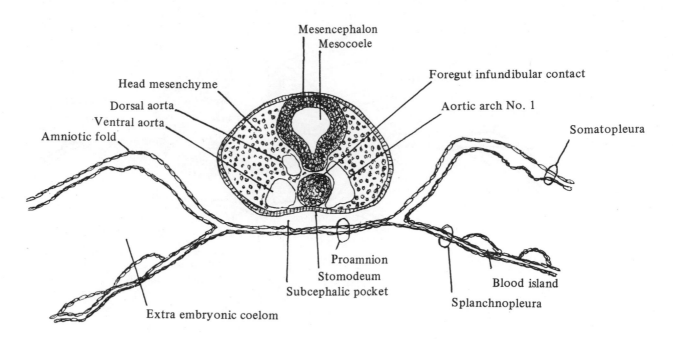

Mesencephalon
Mesocoele
Head mesenchyme
Dorsal aorta
Ventral aorta
Amniotic fold
Foregut infundibular contact
Aortic arch No. 1
Somatopleura
Proamnion
Stomodeum
Subcephalic pocket
Blood island
Splanchnopleura
Extra embryonic coelom

Section 3

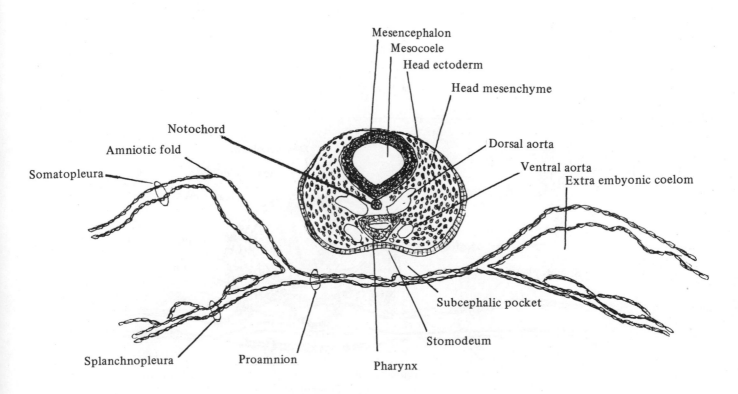

Mesencephalon
Mesocoele
Head ectoderm
Head mesenchyme
Notochord
Amniotic fold
Dorsal aorta
Somatopleura
Ventral aorta
Extra embyonic coelom
Subcephalic pocket
Stomodeum
Splanchnopleura
Proamnion
Pharynx

Section 4

Development of the chick embryo IV
x 100 x. sec.

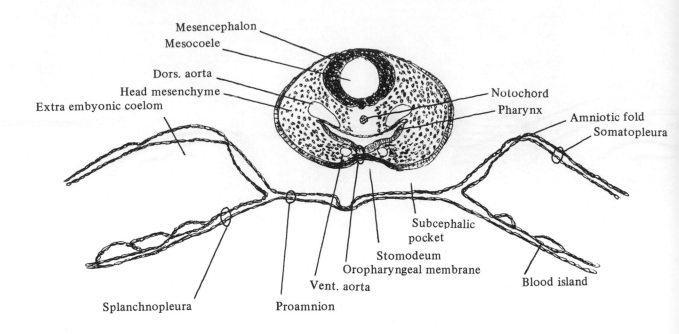

Mesencephalon
Mesocoele
Dors. aorta
Head mesenchyme
Extra embyonic coelom
Notochord
Pharynx
Amniotic fold
Somatopleura
Subcephalic pocket
Stomodeum
Oropharyngeal membrane
Vent. aorta
Blood island
Splanchnopleura
Proamnion

Section 5

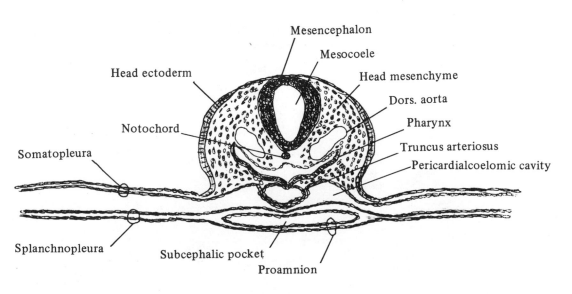

Mesencephalon
Mesocoele
Head ectoderm
Head mesenchyme
Dors. aorta
Notochord
Pharynx
Truncus arteriosus
Somatopleura
Pericardialcoelomic cavity
Splanchnopleura
Subcephalic pocket
Proamnion

Section 6

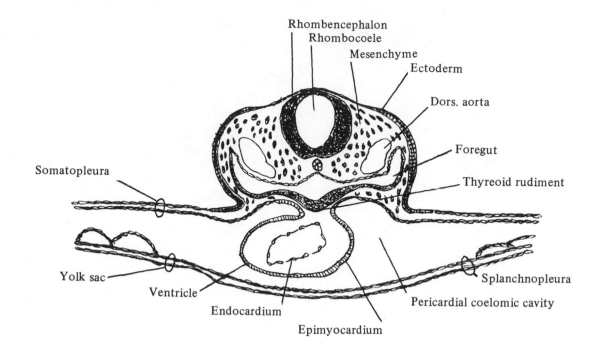

Rhombencephalon
Rhombocoele
Mesenchyme
Ectoderm
Dors. aorta
Foregut
Thyreoid rudiment

Somatopleura

Splanchnopleura

Yolk sac
Ventricle
Endocardium
Epimyocardium
Pericardial coelomic cavity

Section 7

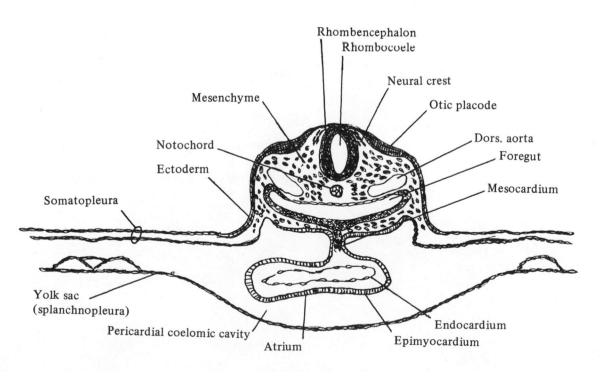

Rhombencephalon
Rhombocoele
Neural crest
Otic placode
Mesenchyme
Dors. aorta
Notochord
Foregut
Ectoderm
Mesocardium
Somatopleura

Yolk sac
(splanchnopleura)
Endocardium
Pericardial coelomic cavity
Epimyocardium
Atrium

Section 8

Development of the chick embryo IV
x 100 x. sec.

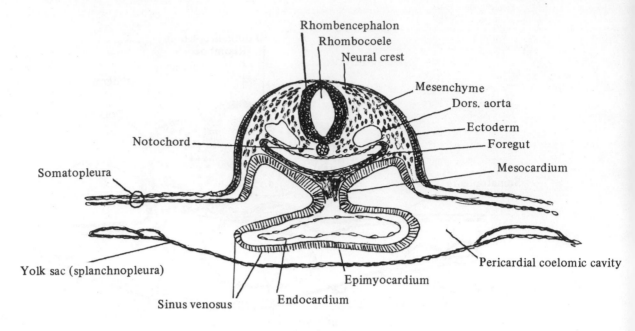

Rhombencephalon
Rhombocoele
Neural crest
Mesenchyme
Dors. aorta
Ectoderm
Foregut
Mesocardium
Notochord
Somatopleura
Yolk sac (splanchnopleura)
Sinus venosus
Endocardium
Epimyocardium
Pericardial coelomic cavity

Section 9

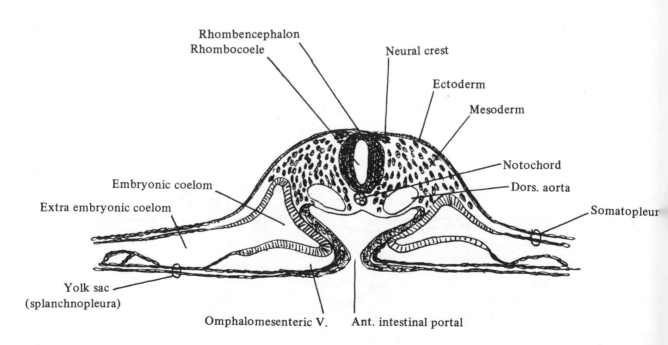

Rhombencephalon
Rhombocoele
Neural crest
Ectoderm
Mesoderm
Embryonic coelom
Extra embryonic coelom
Notochord
Dors. aorta
Somatopleur
Yolk sac
(splanchnopleura)
Omphalomesenteric V.
Ant. intestinal portal

Section 10

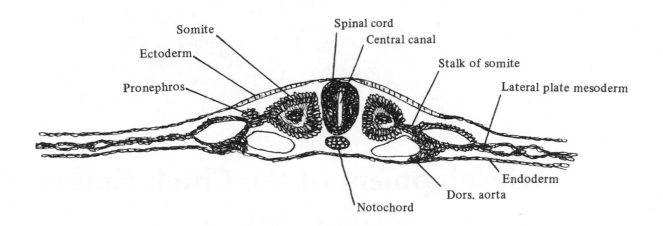

Somite
Spinal cord
Central canal
Ectoderm
Stalk of somite
Pronephros
Lateral plate mesoderm
Endoderm
Dors. aorta
Notochord

Section 11

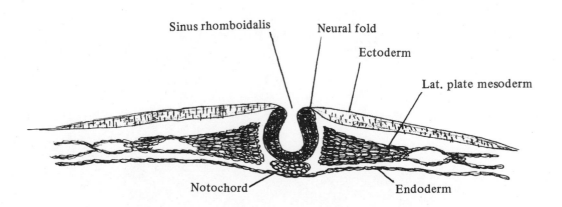

Sinus rhomboidalis
Neural fold
Ectoderm
Lat. plate mesoderm
Notochord
Endoderm

Section 12

Development of the Chick Embryo V
48-Hour Chick Embryo

Development of the chick embryo V
x 40 w.m.

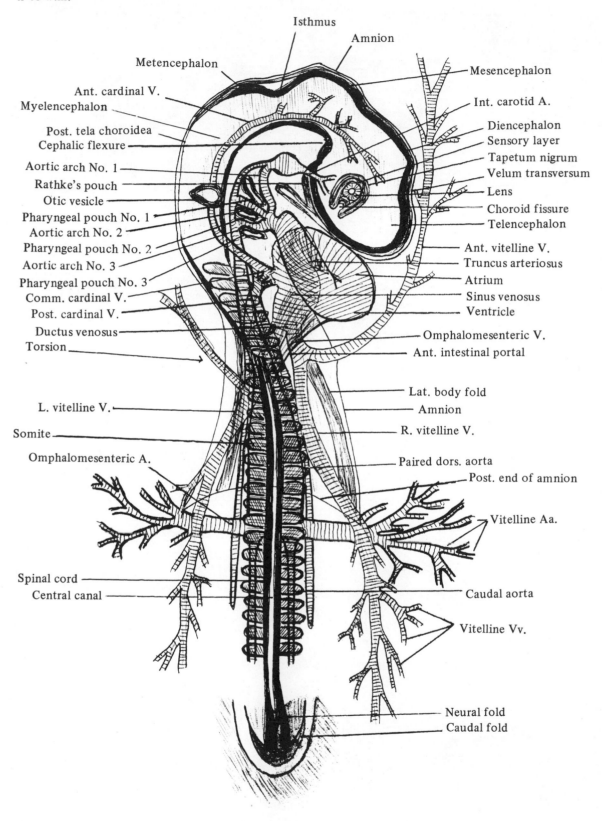

Isthmus
Amnion
Metencephalon
Mesencephalon
Ant. cardinal V.
Myelencephalon
Int. carotid A.
Post. tela choroidea
Diencephalon
Cephalic flexure
Sensory layer
Aortic arch No. 1
Tapetum nigrum
Rathke's pouch
Velum transversum
Otic vesicle
Lens
Pharyngeal pouch No. 1
Choroid fissure
Aortic arch No. 2
Telencephalon
Pharyngeal pouch No. 2
Aortic arch No. 3
Ant. vitelline V.
Pharyngeal pouch No. 3
Truncus arteriosus
Comm. cardinal V.
Atrium
Post. cardinal V.
Sinus venosus
Ductus venosus
Ventricle
Torsion
Omphalomesenteric V.
Ant. intestinal portal
Lat. body fold
L. vitelline V.
Amnion
Somite
R. vitelline V.
Omphalomesenteric A.
Paired dors. aorta
Post. end of amnion
Vitelline Aa.
Spinal cord
Central canal
Caudal aorta
Vitelline Vv.
Neural fold
Caudal fold

Development of the chick embryo V
x 40 w.m.

Sections.

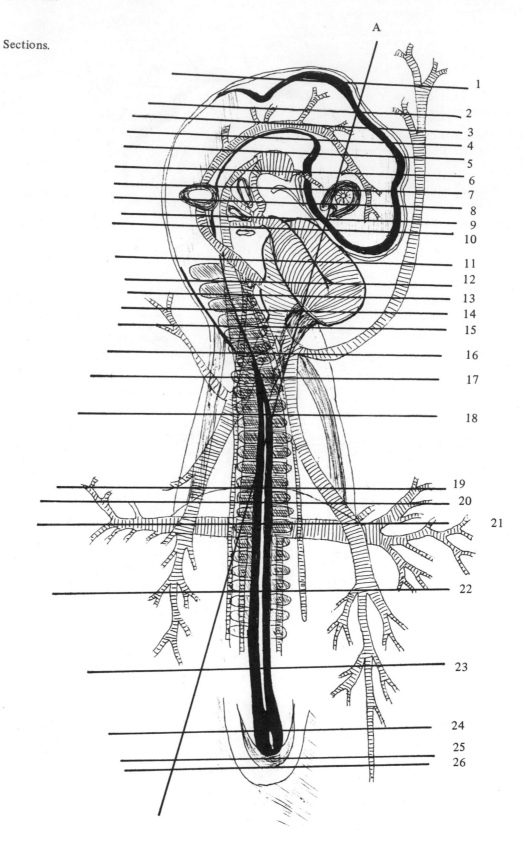

A

1
2
3
4
5
6
7
8
9
10
11
12
13
14
15
16
17
18
19
20
21
22
23
24
25
26

Development of the chick embryo V
x 40 frontal sec.

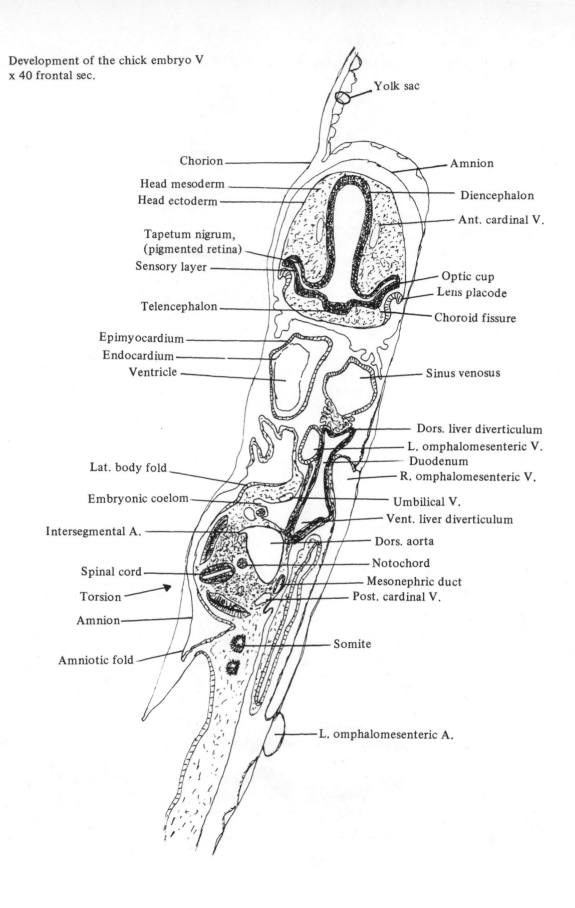

Yolk sac

Chorion

Amnion

Head mesoderm

Diencephalon

Head ectoderm

Ant. cardinal V.

Tapetum nigrum,
(pigmented retina)

Sensory layer

Optic cup

Lens placode

Telencephalon

Choroid fissure

Epimyocardium

Endocardium

Ventricle

Sinus venosus

Dors. liver diverticulum

L. omphalomesenteric V.

Duodenum

Lat. body fold

R. omphalomesenteric V.

Embryonic coelom

Umbilical V.

Vent. liver diverticulum

Intersegmental A.

Dors. aorta

Notochord

Spinal cord

Mesonephric duct

Torsion

Post. cardinal V.

Amnion

Somite

Amniotic fold

L. omphalomesenteric A.

Development of the chick embryo V
x 40 x. sec.

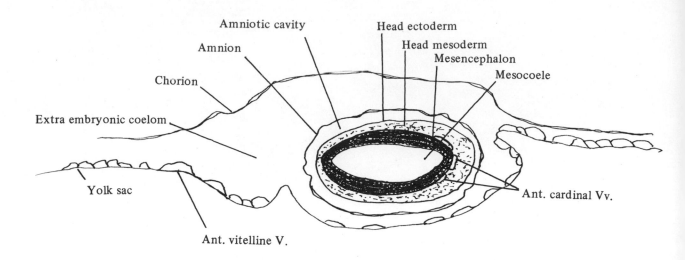

Amniotic cavity

Amnion

Chorion

Extra embryonic coelom

Head ectoderm

Head mesoderm

Mesencephalon

Mesocoele

Ant. cardinal Vv.

Yolk sac

Ant. vitelline V.

Section 1

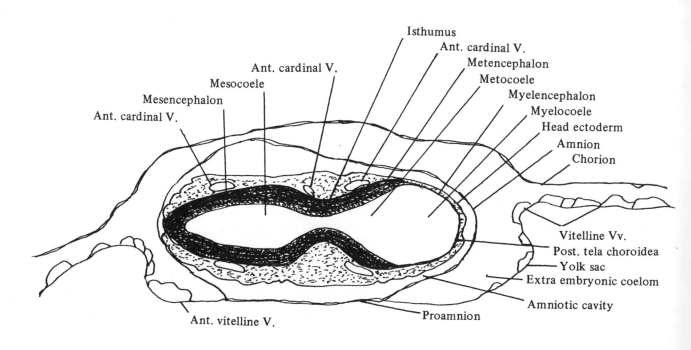

Isthumus

Ant. cardinal V.

Metencephalon

Metocoele

Myelencephalon

Myelocoele

Head ectoderm

Amnion

Chorion

Ant. cardinal V.

Mesocoele

Mesencephalon

Ant. cardinal V.

Vitelline Vv.

Post. tela choroidea

Yolk sac

Extra embryonic coelom

Amniotic cavity

Ant. vitelline V.

Proamnion

Section 2

144

Development of the chick embryo V
x 40 x. sec.

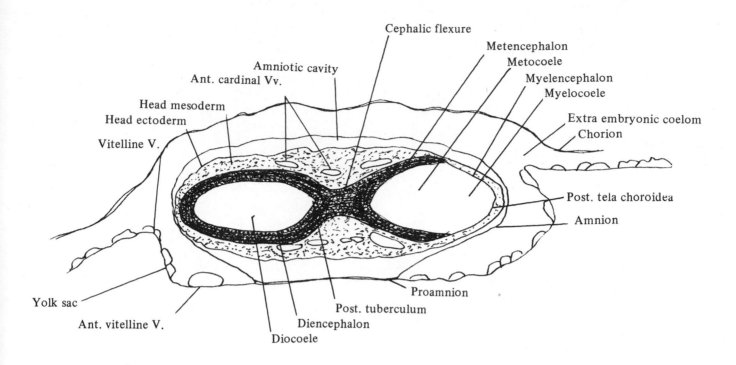

Cephalic flexure

Metencephalon
Metocoele
Myelencephalon
Myelocoele

Amniotic cavity
Ant. cardinal Vv.

Head mesoderm
Head ectoderm

Extra embryonic coelom
Chorion

Vitelline V.

Post. tela choroidea

Amnion

Yolk sac

Proamnion

Ant. vitelline V.

Post. tuberculum
Diencephalon

Diocoele

Section 3

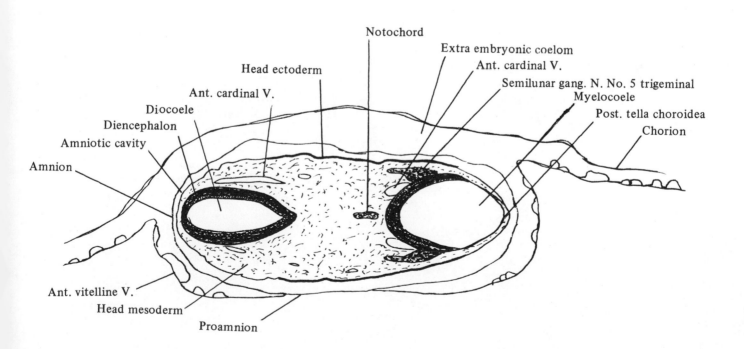

Notochord

Extra embryonic coelom
Ant. cardinal V.

Head ectoderm

Semilunar gang. N. No. 5 trigeminal
Myelocoele

Ant. cardinal V.

Post. tela choroidea
Chorion

Diocoele
Diencephalon

Amniotic cavity

Amnion

Ant. vitelline V.

Head mesoderm

Proamnion

Section 4

Development of the chick embryo V
x 40 x. sec.

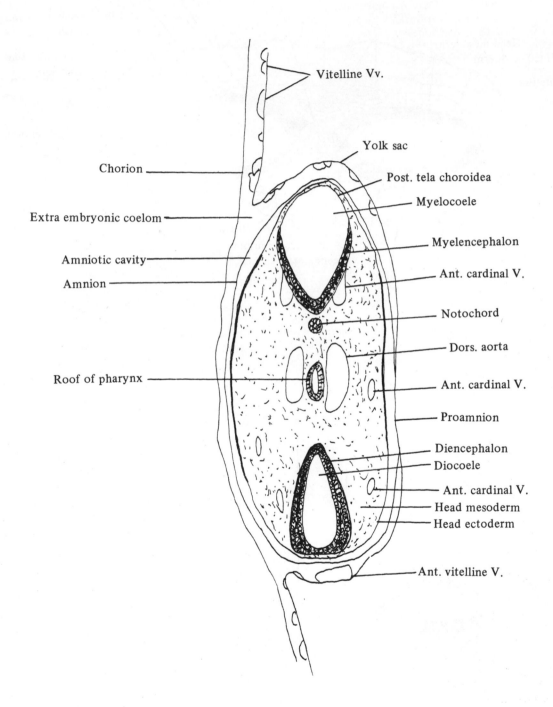

Vitelline Vv.

Chorion

Extra embryonic coelom

Amniotic cavity

Amnion

Roof of pharynx

Yolk sac

Post. tela choroidea

Myelocoele

Myelencephalon

Ant. cardinal V.

Notochord

Dors. aorta

Ant. cardinal V.

Proamnion

Diencephalon
Diocoele

Ant. cardinal V.
Head mesoderm
Head ectoderm

Ant. vitelline V.

Section 5

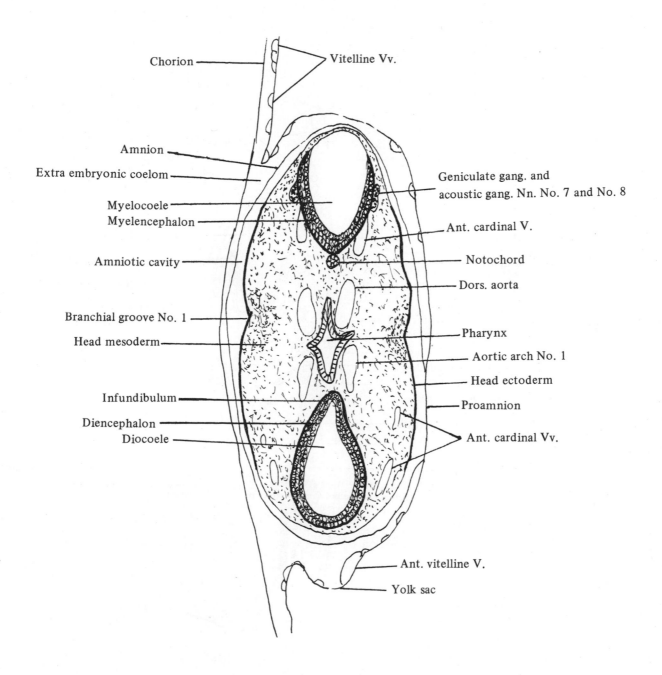

Chorion —————— Vitelline Vv.

Amnion ————
Extra embryonic coelom ————
Geniculate gang. and
acoustic gang. Nn. No. 7 and No. 8

Myelocoele ————
Myelencephalon ————
Ant. cardinal V.

Amniotic cavity ————
Notochord

Dors. aorta

Branchial groove No. 1 ————
Head mesoderm ————
Pharynx
Aortic arch No. 1

Head ectoderm

Infundibulum ————
Proamnion

Diencephalon ————
Diocoele ————
Ant. cardinal Vv.

Ant. vitelline V.
Yolk sac

Section 6

Development of the chick embryo V
x 40 x. sec.

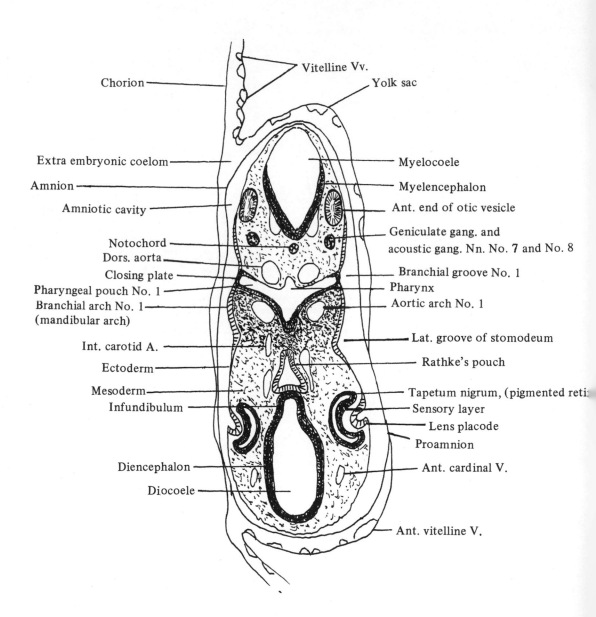

Chorion

Vitelline Vv.

Yolk sac

Extra embryonic coelom

Myelocoele

Amnion

Myelencephalon

Amniotic cavity

Ant. end of otic vesicle

Notochord

Geniculate gang. and
acoustic gang. Nn. No. 7 and No. 8

Dors. aorta

Closing plate

Branchial groove No. 1

Pharyngeal pouch No. 1

Pharynx

Branchial arch No. 1
(mandibular arch)

Aortic arch No. 1

Int. carotid A.

Lat. groove of stomodeum

Ectoderm

Rathke's pouch

Mesoderm

Tapetum nigrum, (pigmented reti

Infundibulum

Sensory layer

Lens placode

Proamnion

Diencephalon

Ant. cardinal V.

Diocoele

Ant. vitelline V.

Section 7

Development of the chick embryo V
x 40 x. sec.

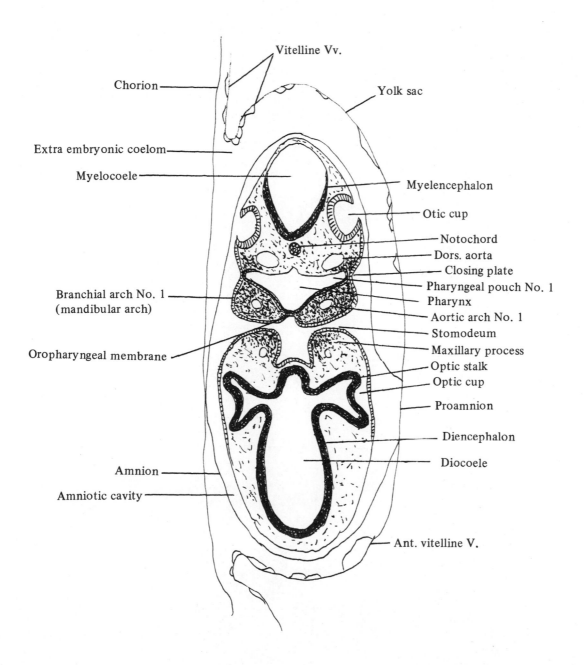

Vitelline Vv.

Chorion

Yolk sac

Extra embryonic coelom

Myelocoele

Myelencephalon

Otic cup

Notochord

Dors. aorta

Closing plate

Branchial arch No. 1
(mandibular arch)

Pharyngeal pouch No. 1

Pharynx

Aortic arch No. 1

Stomodeum

Maxillary process

Oropharyngeal membrane

Optic stalk

Optic cup

Proamnion

Diencephalon

Diocoele

Amnion

Amniotic cavity

Ant. vitelline V.

Section 8

Development of the chick embryo V
x 40 x. sec.

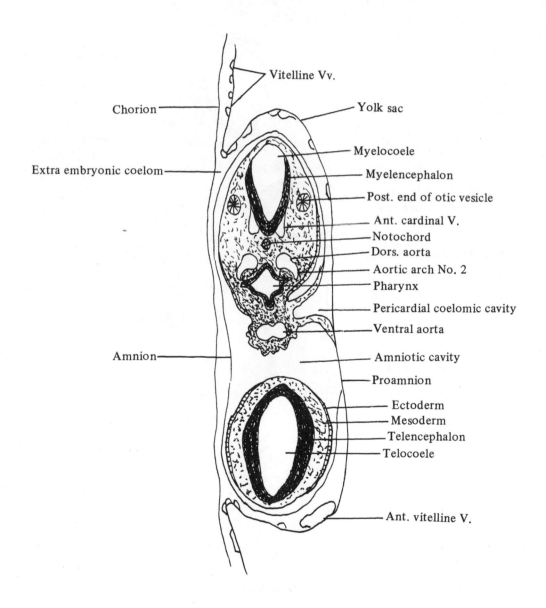

Vitelline Vv.

Chorion

Yolk sac

Extra embryonic coelom

Myelocoele

Myelencephalon

Post. end of otic vesicle

Ant. cardinal V.

Notochord

Dors. aorta

Aortic arch No. 2

Pharynx

Pericardial coelomic cavity

Ventral aorta

Amnion

Amniotic cavity

Proamnion

Ectoderm

Mesoderm

Telencephalon

Telocoele

Ant. vitelline V.

Section 9

Development of the chick embryo V
x 40 x. sec.

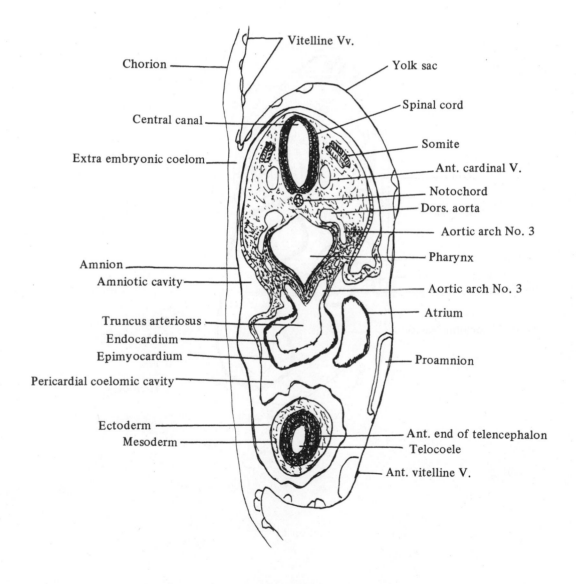

Vitelline Vv.

Chorion

Yolk sac

Central canal

Spinal cord

Extra embryonic coelom

Somite

Ant. cardinal V.

Notochord

Dors. aorta

Aortic arch No. 3

Amnion

Pharynx

Amniotic cavity

Aortic arch No. 3

Truncus arteriosus

Atrium

Endocardium

Epimyocardium

Proamnion

Pericardial coelomic cavity

Ectoderm

Mesoderm

Ant. end of telencephalon

Telocoele

Ant. vitelline V.

Section 10

Development of the chick embry V
x 40 x. sec.

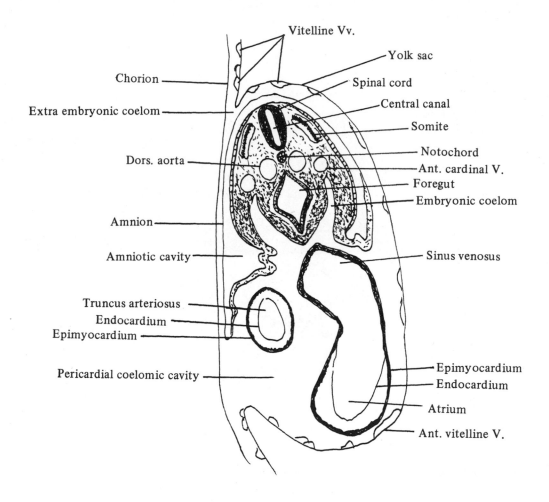

Vitelline Vv.

Yolk sac

Chorion

Spinal cord

Extra embryonic coelom

Central canal

Somite

Dors. aorta

Notochord

Ant. cardinal V.

Foregut

Embryonic coelom

Amnion

Amniotic cavity

Sinus venosus

Truncus arteriosus

Endocardium

Epimyocardium

Pericardial coelomic cavity

Epimyocardium

Endocardium

Atrium

Ant. vitelline V.

Section 11

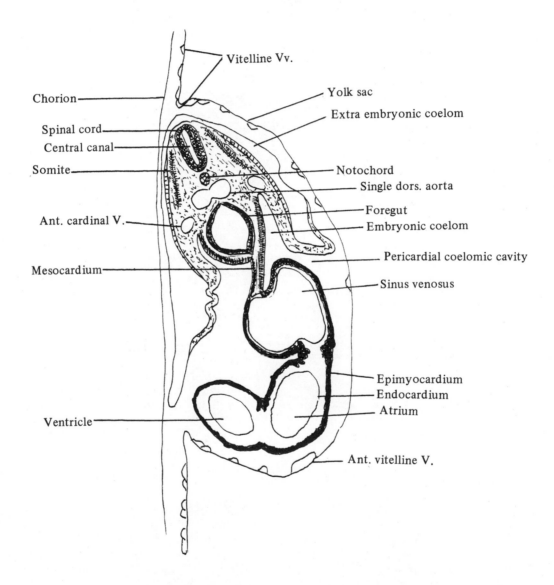

Vitelline Vv.

Chorion

Yolk sac

Extra embryonic coelom

Spinal cord

Central canal

Somite

Notochord

Single dors. aorta

Ant. cardinal V.

Foregut

Embryonic coelom

Pericardial coelomic cavity

Mesocardium

Sinus venosus

Epimyocardium

Endocardium

Atrium

Ventricle

Ant. vitelline V.

Section 12

Development of the chick embryo V
x 40 x. sec.

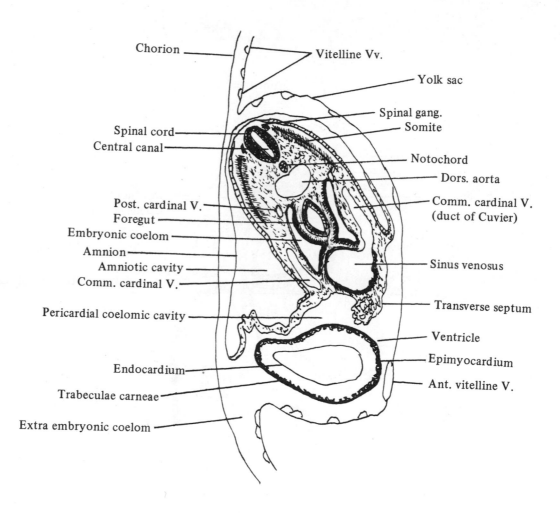

Chorion — Vitelline Vv.

Yolk sac

Spinal gang.

Somite

Spinal cord —
Central canal —

Notochord

Dors. aorta

Post. cardinal V. —
Foregut —
Embryonic coelom —
Amnion —
Amniotic cavity —
Comm. cardinal V. —

Comm. cardinal V.
(duct of Cuvier)

Sinus venosus

Pericardial coelomic cavity —

Transverse septum

Ventricle

Epimyocardium

Endocardium —

Ant. vitelline V.

Trabeculae carneae —

Extra embryonic coelom —

Section 13

Development of the chick embryo V
x 40 x. sec.

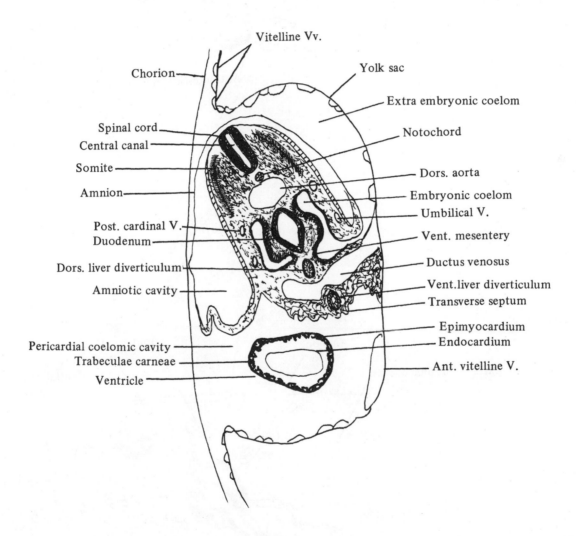

Vitelline Vv.

Chorion

Yolk sac

Extra embryonic coelom

Spinal cord

Notochord

Central canal

Somite

Dors. aorta

Amnion

Embryonic coelom

Umbilical V.

Post. cardinal V.

Duodenum

Vent. mesentery

Dors. liver diverticulum

Ductus venosus

Amniotic cavity

Vent. liver diverticulum

Transverse septum

Epimyocardium

Pericardial coelomic cavity

Endocardium

Trabeculae carneae

Ant. vitelline V.

Ventricle

Section 14

Development of the chick embryo V
x 40 x. sec.

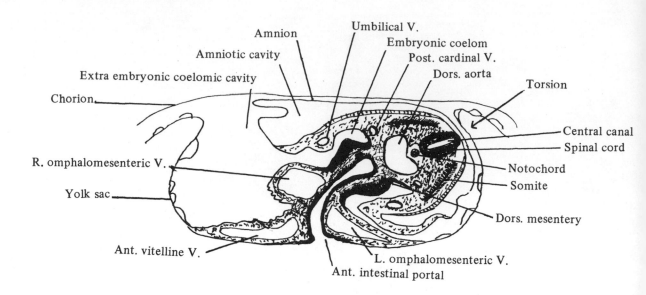

Amnion
Amniotic cavity
Extra embryonic coelomic cavity
Chorion
Umbilical V.
Embryonic coelom
Post. cardinal V.
Dors. aorta
Torsion
Central canal
Spinal cord
R. omphalomesenteric V.
Notochord
Somite
Yolk sac
Dors. mesentery
Ant. vitelline V.
L. omphalomesenteric V.
Ant. intestinal portal

Section 15

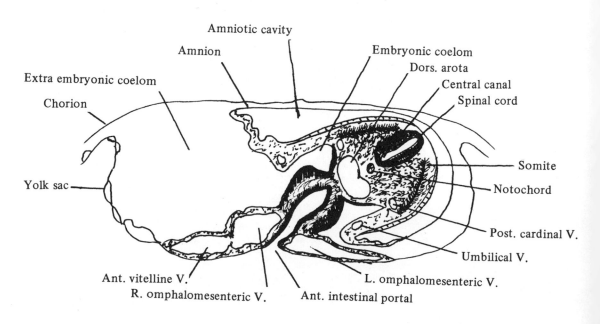

Amniotic cavity
Amnion
Extra embryonic coelom
Embryonic coelom
Dors. arota
Central canal
Spinal cord
Chorion
Yolk sac
Somite
Notochord
Post. cardinal V.
Umbilical V.
Ant. vitelline V.
R. omphalomesenteric V.
L. omphalomesenteric V.
Ant. intestinal portal

Section 16

156

Development of the chick embryo V
x 40 x. sec.

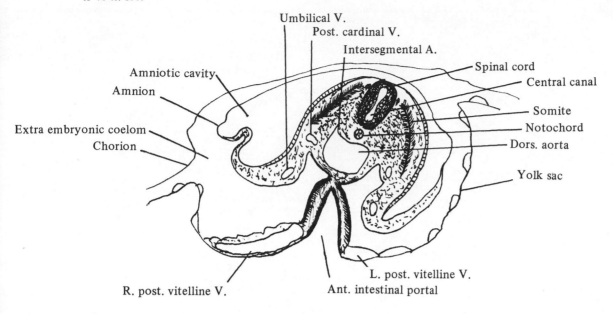

Umbilical V.
Post. cardinal V.
Intersegmental A.
Amniotic cavity
Amnion
Extra embryonic coelom
Chorion
Spinal cord
Central canal
Somite
Notochord
Dors. aorta
Yolk sac
R. post. vitelline V.
Ant. intestinal portal
L. post. vitelline V.

Section 17

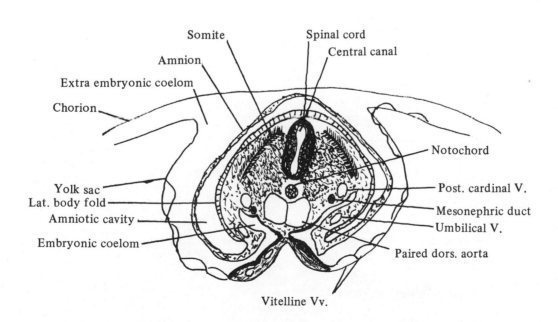

Somite
Amnion
Extra embryonic coelom
Chorion
Spinal cord
Central canal
Notochord
Yolk sac
Lat. body fold
Amniotic cavity
Embryonic coelom
Post. cardinal V.
Mesonephric duct
Umbilical V.
Paired dors. aorta
Vitelline Vv.

Section 18

Development of the chick embryo V
x 40 x. sec.

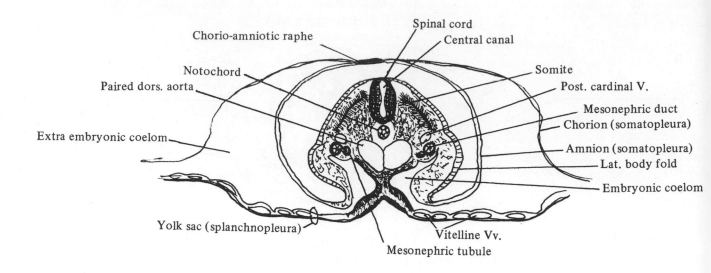

Chorio-amniotic raphe   Spinal cord
Central canal

Notochord   Somite
Paired dors. aorta   Post. cardinal V.

Mesonephric duct
Chorion (somatopleura)
Extra embryonic coelom

Amnion (somatopleura)
Lat. body fold

Embryonic coelom

Yolk sac (splanchnopleura)   Vitelline Vv.
Mesonephric tubule

Section 19

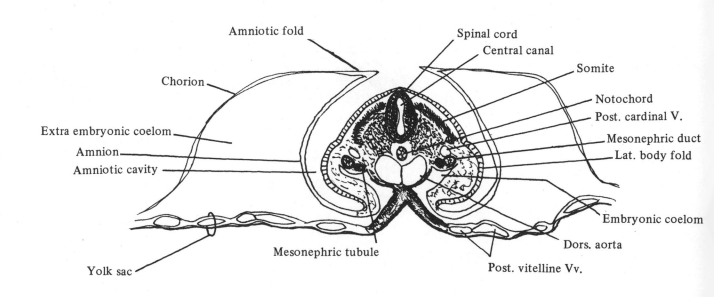

Amniotic fold   Spinal cord
Central canal

Chorion   Somite

Notochord
Extra embryonic coelom   Post. cardinal V.

Amnion   Mesonephric duct
Amniotic cavity   Lat. body fold

Embryonic coelom

Dors. aorta
Mesonephric tubule
Post. vitelline Vv.
Yolk sac

Section 20

Development of the chick embryo V
x 40 x. sec.

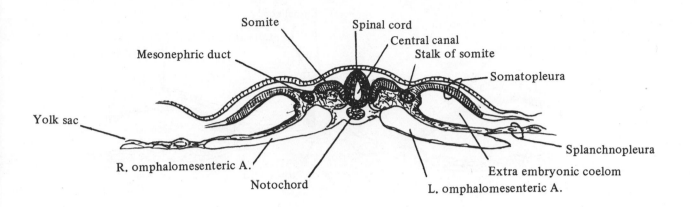

Somite   Spinal cord
Central canal
Mesonephric duct   Stalk of somite
Somatopleura
Yolk sac
R. omphalomesenteric A.   Splanchnopleura
Notochord   Extra embryonic coelom
L. omphalomesenteric A.

Section 21

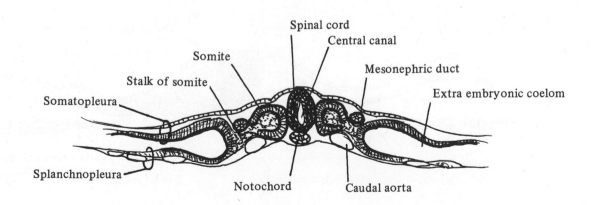

Spinal cord
Central canal
Somite
Stalk of somite   Mesonephric duct
Somatopleura   Extra embryonic coelom
Splanchnopleura
Notochord   Caudal aorta

Section 22

Development of the chick embryo V
x 40 x. sec.

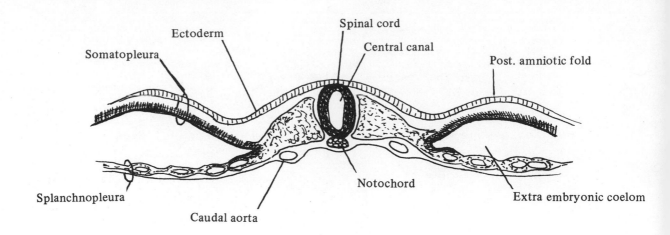

Section 23

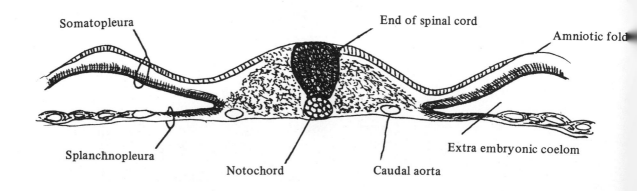

Section 24

Development of the chick embryo V
x 40 x. sec.

Somatopleura

Ectoderm

Post. amniotic fold

Extra embryonic coelom

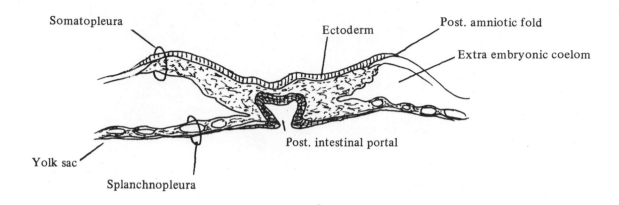

Yolk sac

Splanchnopleura

Post. intestinal portal

Section 25

Mesoderm

Ectoderm

Hindgut

Extra embryonic coelom

Somatopleura

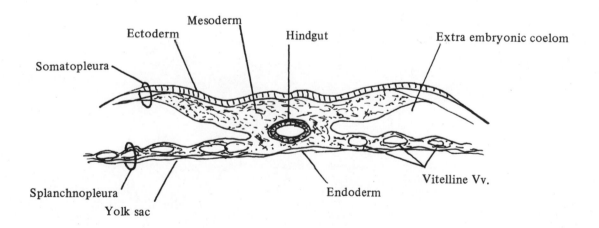

Splanchnopleura

Yolk sac

Vitelline Vv.

Endoderm

Section 26

Development of the Chick Embryo VI
72-Hour Chick Embryo

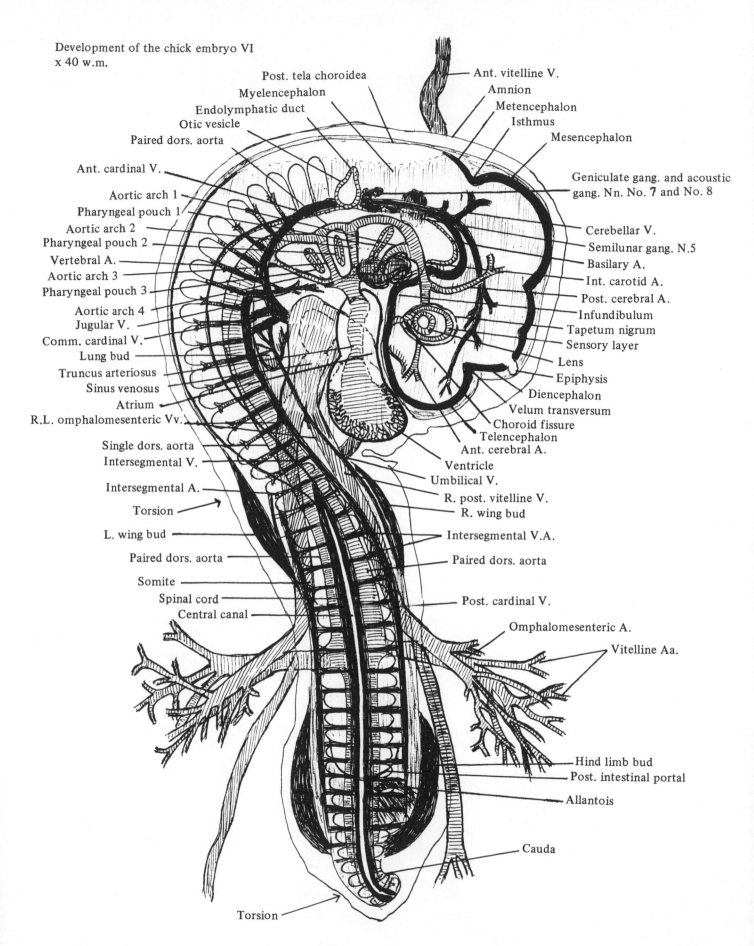

Development of the chick embryo VI
x 40 w.m.

Post. tela choroidea
Myelencephalon
Endolymphatic duct
Otic vesicle
Paired dors. aorta

Ant. vitelline V.
Amnion
Metencephalon
Isthmus
Mesencephalon

Ant. cardinal V.
Aortic arch 1
Pharyngeal pouch 1
Aortic arch 2
Pharyngeal pouch 2
Vertebral A.
Aortic arch 3
Pharyngeal pouch 3
Aortic arch 4
Jugular V.
Comm. cardinal V.
Lung bud
Truncus arteriosus
Sinus venosus
Atrium
R.L. omphalomesenteric Vv.

Geniculate gang. and acoustic
gang. Nn. No. 7 and No. 8
Cerebellar V.
Semilunar gang. N.5
Basilary A.
Int. carotid A.
Post. cerebral A.
Infundibulum
Tapetum nigrum
Sensory layer
Lens
Epiphysis
Diencephalon
Velum transversum
Choroid fissure
Telencephalon
Ant. cerebral A.

Single dors. aorta
Intersegmental V.
Intersegmental A.
Torsion
L. wing bud
Paired dors. aorta
Somite
Spinal cord
Central canal

Ventricle
Umbilical V.
R. post. vitelline V.
R. wing bud
Intersegmental V.A.
Paired dors. aorta
Post. cardinal V.

Omphalomesenteric A.
Vitelline Aa.

Hind limb bud
Post. intestinal portal
Allantois

Cauda

Torsion

163

Development of the chick embryo VI
x 100 w.m.

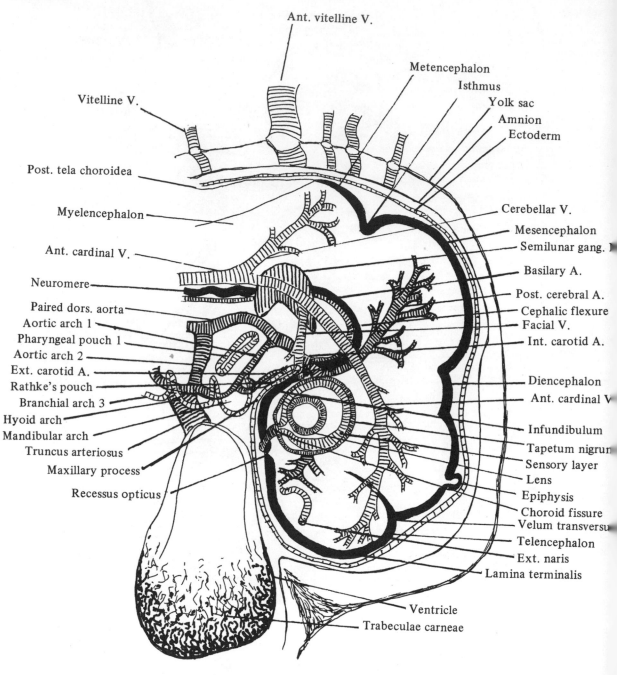

Ant. vitelline V.

Metencephalon
Isthmus
Yolk sac
Amnion
Ectoderm

Vitelline V.

Post. tela choroidea

Myelencephalon

Cerebellar V.
Mesencephalon
Semilunar gang.
Basilary A.

Ant. cardinal V.

Neuromere
Post. cerebral A.
Cephalic flexure
Paired dors. aorta
Facial V.
Aortic arch 1
Int. carotid A.
Pharyngeal pouch 1
Aortic arch 2
Ext. carotid A.
Rathke's pouch
Diencephalon
Branchial arch 3
Ant. cardinal V
Hyoid arch
Mandibular arch
Infundibulum
Truncus arteriosus
Tapetum nigrum
Sensory layer
Maxillary process
Lens
Recessus opticus
Epiphysis
Choroid fissure
Velum transversum
Telencephalon
Ext. naris
Lamina terminalis

Ventricle
Trabeculae carneae

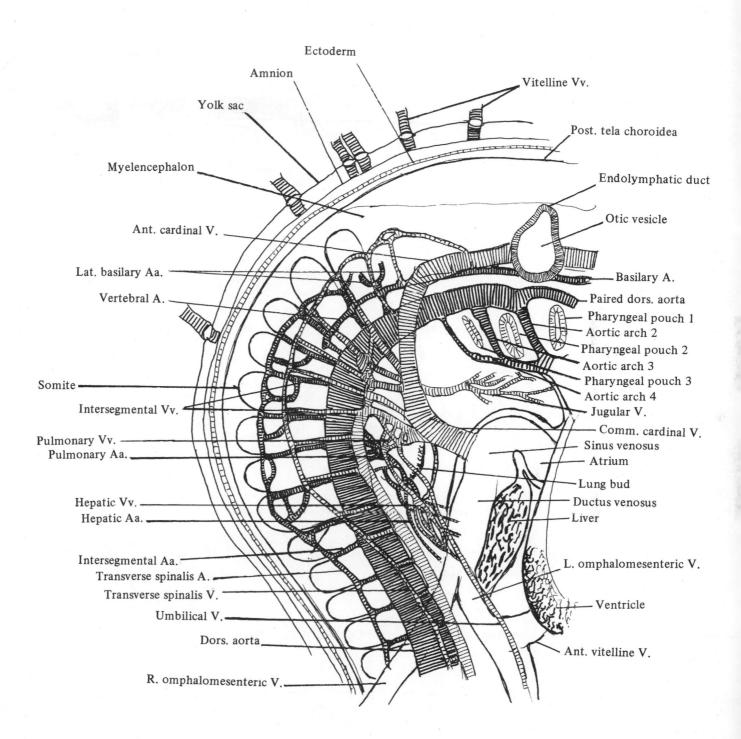

Ectoderm

Amnion

Yolk sac

Vitelline Vv.

Post. tela choroidea

Myelencephalon

Endolymphatic duct

Otic vesicle

Ant. cardinal V.

Lat. basilary Aa.

Basilary A.

Vertebral A.

Paired dors. aorta

Pharyngeal pouch 1

Aortic arch 2

Pharyngeal pouch 2

Aortic arch 3

Pharyngeal pouch 3

Aortic arch 4

Somite

Jugular V.

Intersegmental Vv.

Comm. cardinal V.

Pulmonary Vv.

Sinus venosus

Pulmonary Aa.

Atrium

Lung bud

Ductus venosus

Hepatic Vv.

Liver

Hepatic Aa.

Intersegmental Aa.

L. omphalomesenteric V.

Transverse spinalis A.

Transverse spinalis V.

Umbilical V.

Ventricle

Dors. aorta

Ant. vitelline V.

R. omphalomesenteric V.

Development of the chick embryo VI
x 40 w.m.

Sections

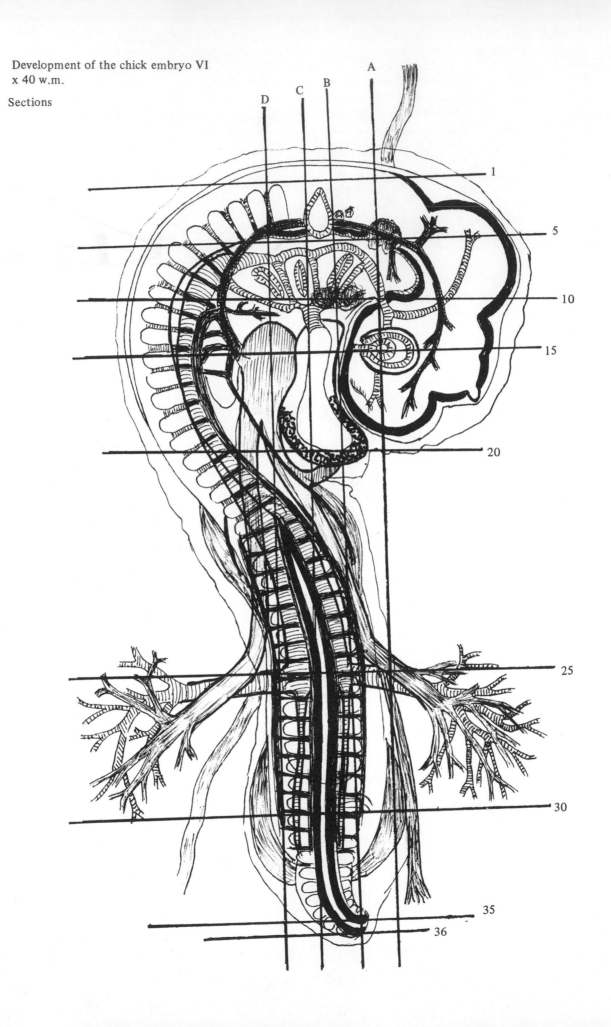

D   C B   A

1

5

10

15

20

25

30

35
36

Development of the chick embryo VI
x 40 frontal sec.

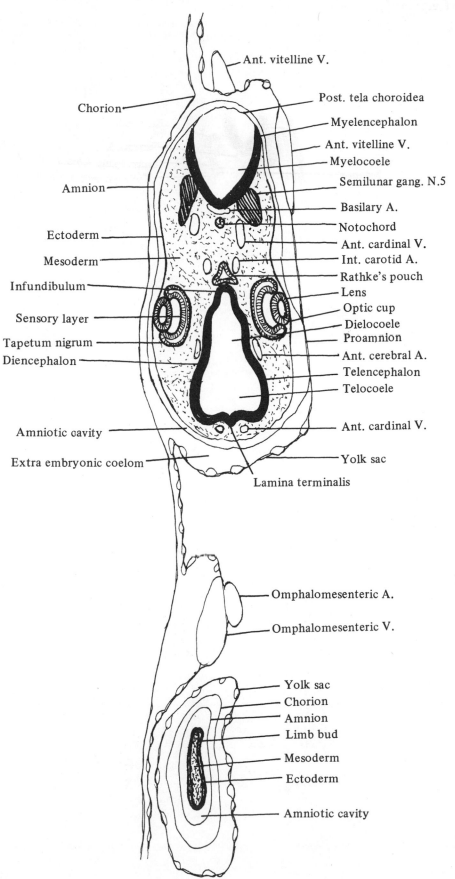

Ant. vitelline V.

Chorion

Post. tela choroidea

Myelencephalon

Ant. vitelline V.

Myelocoele

Amnion

Semilunar gang. N.5

Basilary A.

Ectoderm

Notochord

Ant. cardinal V.

Mesoderm

Int. carotid A.

Rathke's pouch

Infundibulum

Lens

Optic cup

Sensory layer

Dielocoele

Proamnion

Tapetum nigrum

Diencephalon

Ant. cerebral A.

Telencephalon

Telocoele

Amniotic cavity

Ant. cardinal V.

Extra embryonic coelom

Yolk sac

Lamina terminalis

Omphalomesenteric A.

Omphalomesenteric V.

Yolk sac

Chorion

Amnion

Limb bud

Mesoderm

Ectoderm

Amniotic cavity

Section A

Development of the chick embryo VI
x 40 frontal sec.

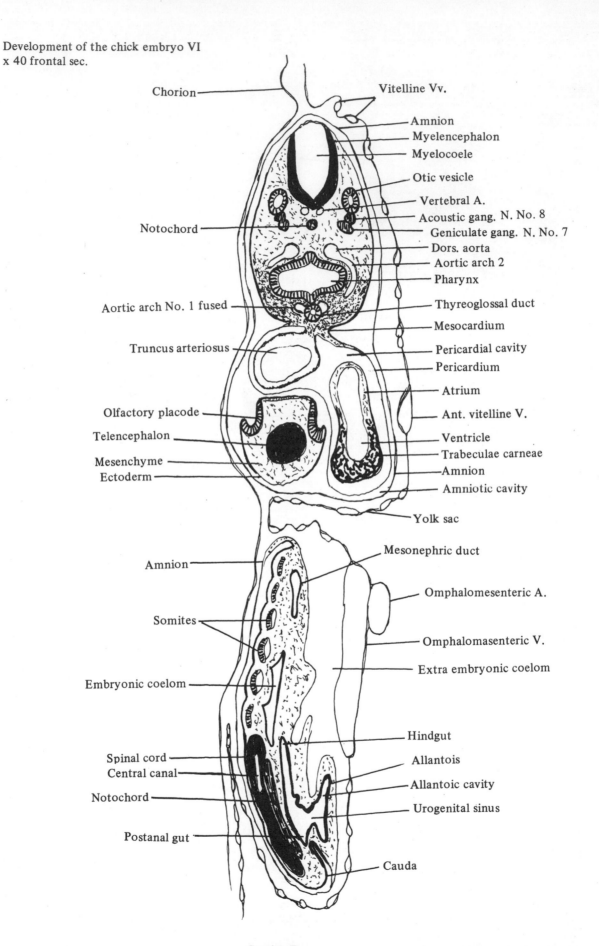

Chorion

Vitelline Vv.

Amnion

Myelencephalon

Myelocoele

Otic vesicle

Vertebral A.

Acoustic gang. N. No. 8

Geniculate gang. N. No. 7

Notochord

Dors. aorta

Aortic arch 2

Pharynx

Thyreoglossal duct

Aortic arch No. 1 fused

Mesocardium

Truncus arteriosus

Pericardial cavity

Pericardium

Atrium

Olfactory placode

Ant. vitelline V.

Telencephalon

Ventricle

Trabeculae carneae

Mesenchyme

Amnion

Ectoderm

Amniotic cavity

Yolk sac

Amnion

Mesonephric duct

Omphalomesenteric A.

Somites

Omphalomasenteric V.

Extra embryonic coelom

Embryonic coelom

Hindgut

Spinal cord

Allantois

Central canal

Allantoic cavity

Notochord

Urogenital sinus

Postanal gut

Cauda

Section B

Development of the chick embryo VI
x 40 frontal sec.

Chorion —————————————— Yolk sac

Amnion ————————

Myelocoele ———— Myelencephalon
Mesoderm

Petrosal gang. N. No. 9

Vertebral A. ———— Post. end of otic vesicle
Ant. cardinal V. ———— Notochord
Dors. aorta ———— Dors. aorta
Pharyngeal pouch No. 3 ———— Pharynx
Branchial arch No. 3 ———— Aortic arch No. 3

Amniotic cavity

Truncus arteriosus ————

Atrium

Endocardium ————

Epimyocardium

Ventricle ————
Trabeculae carneae ———— Ant. vitelline V.

Extra embryonic coelom ————

Umbilical Vv.

Amniotic cavity ————
Amnion ———— L. post. vitelline V.
Post. cardinal V. ———— Mesonephric duct.

Post. intestinal portal

Intersegmental A. ———— Dors. aorta
L. post. vitelline V.
Somite ———— Embryonic coelom
Hindgut ———— Allantoic vessels
Spinal cord ———— Allantoic cavity
Central canal ———— Cloaca
Notochord ————

Postanal gut

Caudal aorta ————
Intersegmental A. ————

Somite ————

Ectoderm ———— Cauda

Section C

169

Development of the chick embryo VI
x 40 frontal sec.

Yolk sac
Chorion
Amnion
Spinal cord
Spinal gang.
Central canal
Somite
Intersegmental A.
Notochord
Dors. aorta
Ant. cardinal V.
Duodenum
Comm. cardinal
Embryonic coelom
Dors. mesentery
Liver
Amniotic cavity
Ductus venosus
Hepatic cords
Extra embryonic coelom
Amnion
L. omphalomesenteric V.
Embryonic coelom
R. omphalomesenteric V.
Umbilical V.
Ant. intestinal portal
Wing bud
Post. vitelline V.
Mesonephric duct
Glomerulus
Post. cardinal V.
Subcardinal V.
Intersegmental A.
Renal A.
Spinal cord
Dors. aorta
Central canal
Notochord
Dors. aorta
Somites
R. omphalomesenteric A.
Intersegmental V.
Post. cardinal V.
Mesonephric duct
Intersegmental V.
Umbilical Vv.
Lat. body fold
Amniotic cavity

Section D

170

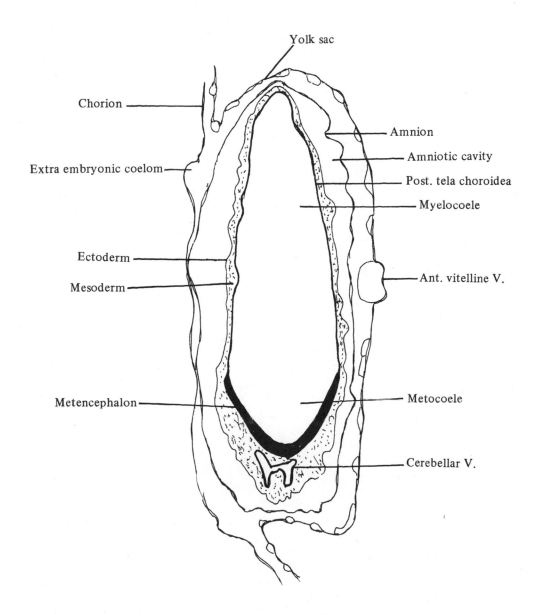

Yolk sac

Chorion

Amnion

Amniotic cavity

Extra embryonic coelom

Post. tela choroidea

Myelocoele

Ectoderm

Mesoderm

Ant. vitelline V.

Metencephalon

Metocoele

Cerebellar V.

Section 1

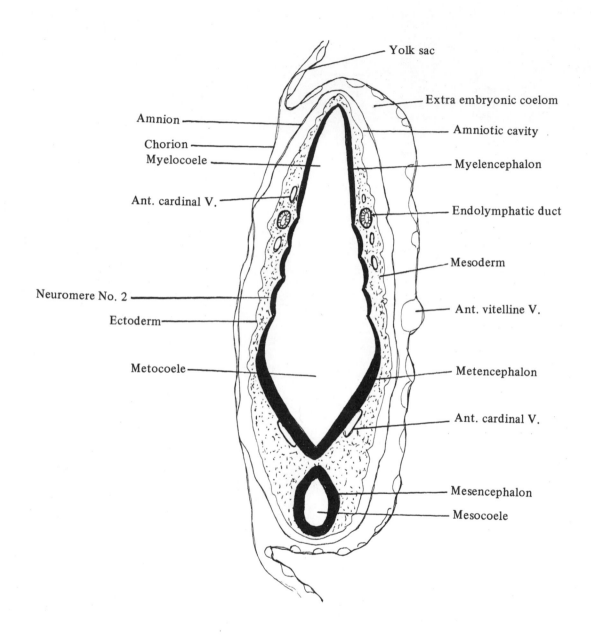

Yolk sac

Extra embryonic coelom

Amnion

Amniotic cavity

Chorion

Myelocoele

Myelencephalon

Ant. cardinal V.

Endolymphatic duct

Mesoderm

Neuromere No. 2

Ant. vitelline V.

Ectoderm

Metocoele

Metencephalon

Ant. cardinal V.

Mesencephalon

Mesocoele

Section 2

Development of the chick embryo VI
x 40 x. sec.

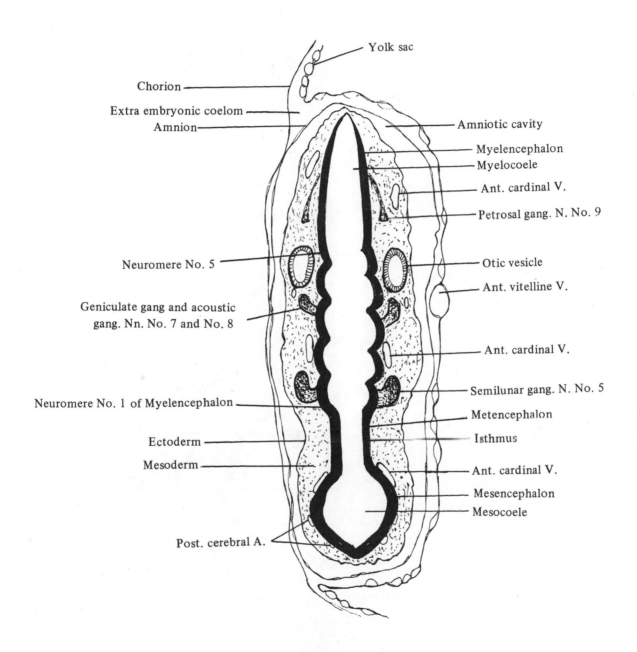

Yolk sac

Chorion

Extra embryonic coelom

Amnion

Amniotic cavity

Myelencephalon

Myelocoele

Ant. cardinal V.

Petrosal gang. N. No. 9

Neuromere No. 5

Otic vesicle

Ant. vitelline V.

Geniculate gang and acoustic
gang. Nn. No. 7 and No. 8

Ant. cardinal V.

Semilunar gang. N. No. 5

Neuromere No. 1 of Myelencephalon

Metencephalon

Ectoderm

Isthmus

Mesoderm

Ant. cardinal V.

Mesencephalon

Mesocoele

Post. cerebral A.

Section 3

173

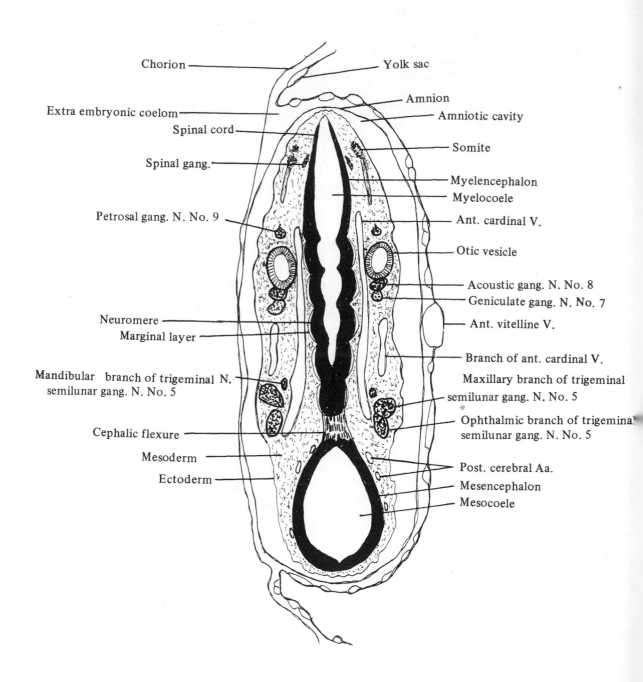

Chorion

Yolk sac

Extra embryonic coelom

Amnion

Amniotic cavity

Spinal cord

Somite

Spinal gang.

Myelencephalon

Myelocoele

Petrosal gang. N. No. 9

Ant. cardinal V.

Otic vesicle

Acoustic gang. N. No. 8

Geniculate gang. N. No. 7

Neuromere

Ant. vitelline V.

Marginal layer

Branch of ant. cardinal V.

Mandibular   branch of trigeminal N.
semilunar gang. N. No. 5

Maxillary branch of trigeminal
semilunar gang. N. No. 5

Ophthalmic branch of trigeminal
semilunar gang. N. No. 5

Cephalic flexure

Mesoderm

Post. cerebral Aa.

Ectoderm

Mesencephalon

Mesocoele

Section 4

174

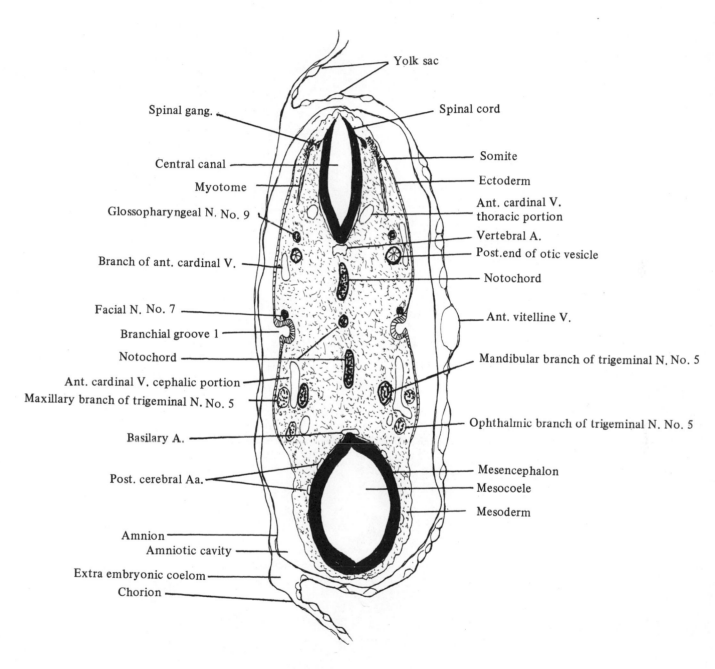

Yolk sac

Spinal gang.

Spinal cord

Central canal

Somite

Myotome

Ectoderm

Glossopharyngeal N. No. 9

Ant. cardinal V.
thoracic portion

Vertebral A.

Branch of ant. cardinal V.

Post. end of otic vesicle

Notochord

Facial N. No. 7

Ant. vitelline V.

Branchial groove 1

Notochord

Mandibular branch of trigeminal N. No. 5

Ant. cardinal V. cephalic portion

Maxillary branch of trigeminal N. No. 5

Basilary A.

Ophthalmic branch of trigeminal N. No. 5

Post. cerebral Aa.

Mesencephalon

Mesocoele

Mesoderm

Amnion

Amniotic cavity

Extra embryonic coelom

Chorion

Section 5

Development of the chick embryo VI
x 40 x. sec.

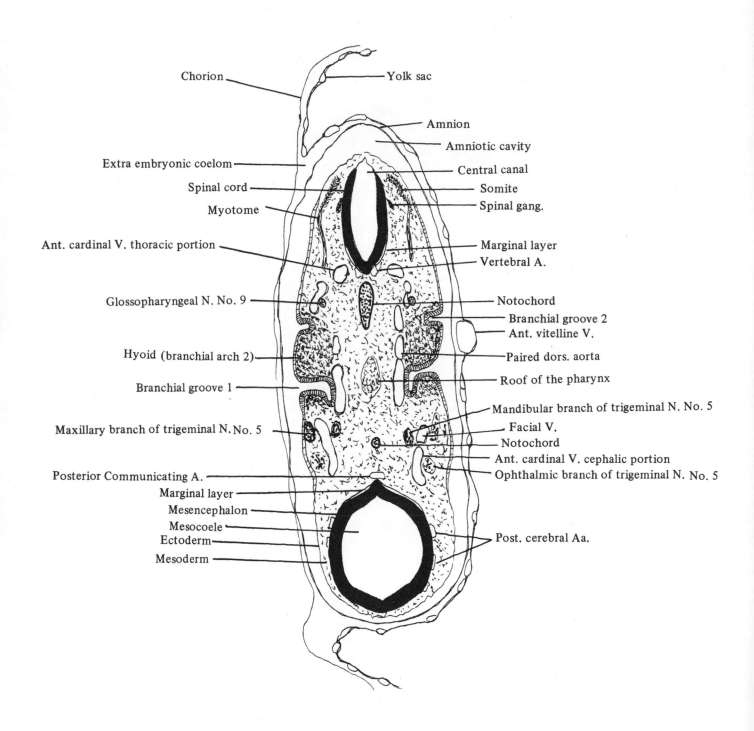

Chorion

Yolk sac

Amnion

Amniotic cavity

Extra embryonic coelom

Central canal

Spinal cord

Somite

Myotome

Spinal gang.

Ant. cardinal V. thoracic portion

Marginal layer

Vertebral A.

Glossopharyngeal N. No. 9

Notochord

Branchial groove 2

Ant. vitelline V.

Hyoid (branchial arch 2)

Paired dors. aorta

Branchial groove 1

Roof of the pharynx

Mandibular branch of trigeminal N. No. 5

Maxillary branch of trigeminal N. No. 5

Facial V.

Notochord

Ant. cardinal V. cephalic portion

Posterior Communicating A.

Ophthalmic branch of trigeminal N. No. 5

Marginal layer

Mesencephalon

Mesocoele

Ectoderm

Post. cerebral Aa.

Mesoderm

Section 6

Development of the chick embryo VI
x 40 x. sec.

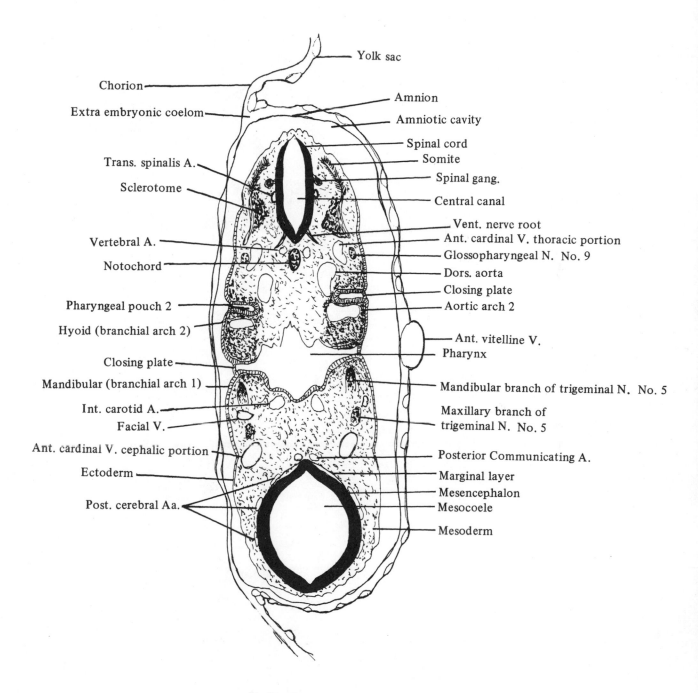

Yolk sac

Chorion

Extra embryonic coelom

Amnion

Amniotic cavity

Spinal cord

Trans. spinalis A.

Somite

Spinal gang.

Sclerotome

Central canal

Vent. nerve root

Vertebral A.

Ant. cardinal V. thoracic portion

Glossopharyngeal N. No. 9

Notochord

Dors. aorta

Closing plate

Pharyngeal pouch 2

Aortic arch 2

Hyoid (branchial arch 2)

Ant. vitelline V.

Pharynx

Closing plate

Mandibular (branchial arch 1)

Mandibular branch of trigeminal N. No. 5

Int. carotid A.

Maxillary branch of
trigeminal N. No. 5

Facial V.

Ant. cardinal V. cephalic portion

Posterior Communicating A.

Ectoderm

Marginal layer

Mesencephalon

Post. cerebral Aa.

Mesocoele

Mesoderm

Section 7

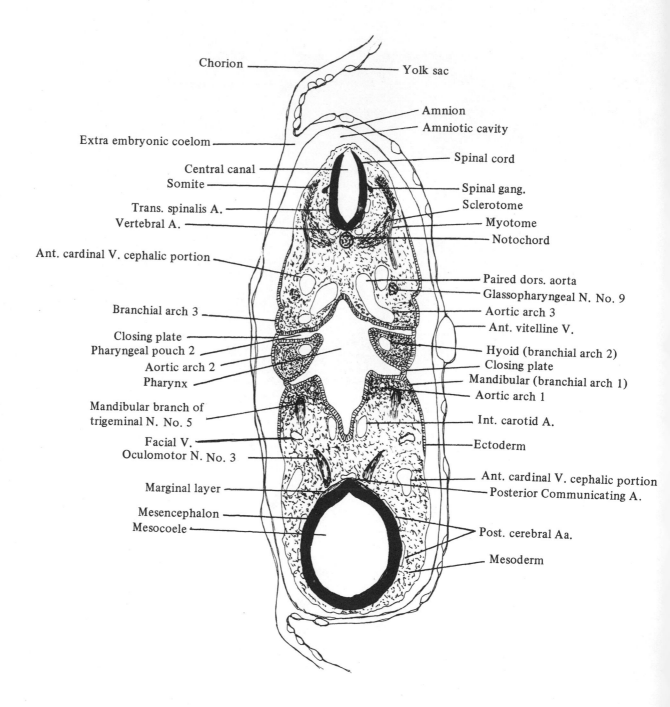

Chorion —————— Yolk sac

Amnion
Amniotic cavity

Extra embryonic coelom —————— Spinal cord

Central canal ——————
Somite ——————

Trans. spinalis A. ——————
Vertebral A. ——————

Spinal gang.
Sclerotome
Myotome
Notochord

Ant. cardinal V. cephalic portion ——————

Paired dors. aorta
Glassopharyngeal N. No. 9

Branchial arch 3 ——————
Aortic arch 3
Ant. vitelline V.

Closing plate ——————
Pharyngeal pouch 2 ——————
Aortic arch 2 ——————
Pharynx ——————

Hyoid (branchial arch 2)
Closing plate
Mandibular (branchial arch 1)
Aortic arch 1

Mandibular branch of
trigeminal N. No. 5 ——————

Int. carotid A.

Facial V. ——————
Oculomotor N. No. 3 ——————

Ectoderm

Marginal layer ——————

Ant. cardinal V. cephalic portion
Posterior Communicating A.

Mesencephalon ——————
Mesocoele ——————

Post. cerebral Aa.

Mesoderm

Section 8

Development of the chick embryo VI
x 40 x. sec.

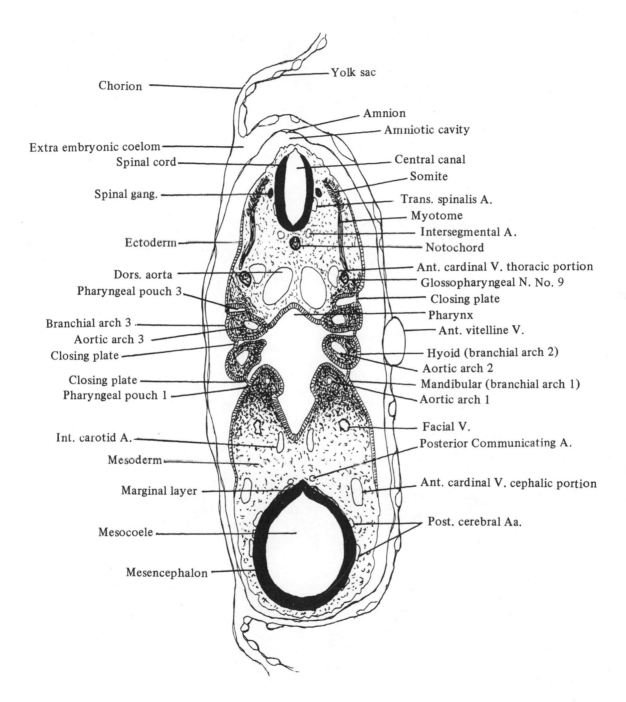

Chorion

Yolk sac

Amnion

Amniotic cavity

Extra embryonic coelom

Central canal

Spinal cord

Somite

Spinal gang.

Trans. spinalis A.

Myotome

Intersegmental A.

Ectoderm

Notochord

Dors. aorta

Ant. cardinal V. thoracic portion

Pharyngeal pouch 3

Glossopharyngeal N. No. 9

Closing plate

Branchial arch 3

Pharynx

Aortic arch 3

Ant. vitelline V.

Closing plate

Hyoid (branchial arch 2)

Aortic arch 2

Closing plate

Mandibular (branchial arch 1)

Pharyngeal pouch 1

Aortic arch 1

Facial V.

Int. carotid A.

Posterior Communicating A.

Mesoderm

Marginal layer

Ant. cardinal V. cephalic portion

Mesocoele

Post. cerebral Aa.

Mesencephalon

Section 9

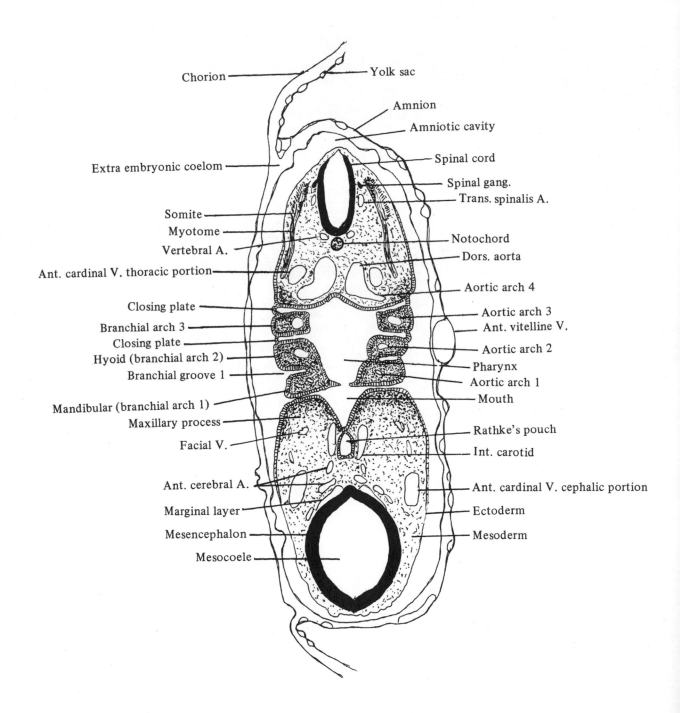

Chorion — Yolk sac

Amnion

Amniotic cavity

Spinal cord

Extra embryonic coelom

Spinal gang.

Trans. spinalis A.

Somite

Myotome

Notochord

Vertebral A.

Dors. aorta

Ant. cardinal V. thoracic portion

Aortic arch 4

Closing plate

Aortic arch 3

Branchial arch 3

Ant. vitelline V.

Closing plate

Hyoid (branchial arch 2)

Aortic arch 2

Branchial groove 1

Pharynx

Aortic arch 1

Mandibular (branchial arch 1)

Mouth

Maxillary process

Rathke's pouch

Facial V.

Int. carotid

Ant. cerebral A.

Ant. cardinal V. cephalic portion

Marginal layer

Ectoderm

Mesencephalon

Mesoderm

Mesocoele

Section 10

Development of the chick embryo VI
x 40 x. sec.

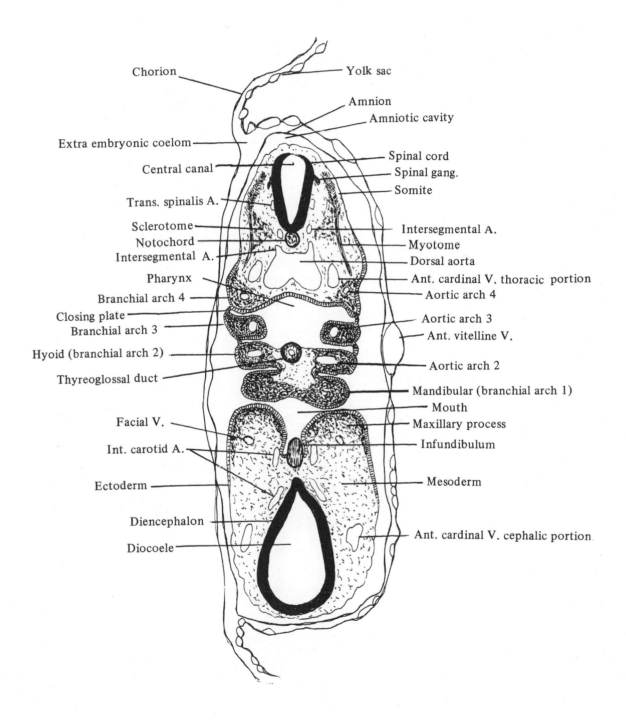

Chorion

Yolk sac

Amnion

Amniotic cavity

Extra embryonic coelom

Spinal cord

Central canal

Spinal gang.

Somite

Trans. spinalis A.

Sclerotome

Intersegmental A.

Notochord

Myotome

Intersegmental A.

Dorsal aorta

Pharynx

Ant. cardinal V. thoracic portion

Branchial arch 4

Aortic arch 4

Closing plate

Aortic arch 3

Branchial arch 3

Ant. vitelline V.

Hyoid (branchial arch 2)

Aortic arch 2

Thyreoglossal duct

Mandibular (branchial arch 1)

Mouth

Facial V.

Maxillary process

Int. carotid A.

Infundibulum

Ectoderm

Mesoderm

Diencephalon

Diocoele

Ant. cardinal V. cephalic portion

Section 11

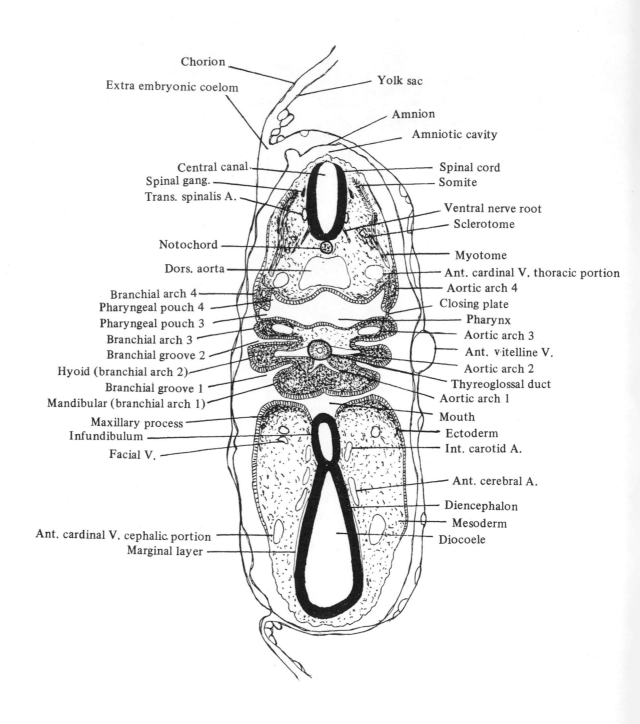

Chorion

Extra embryonic coelom

Yolk sac

Amnion

Amniotic cavity

Central canal

Spinal gang.

Trans. spinalis A.

Spinal cord

Somite

Ventral nerve root

Sclerotome

Notochord

Myotome

Dors. aorta

Ant. cardinal V. thoracic portion

Branchial arch 4

Pharyngeal pouch 4

Pharyngeal pouch 3

Branchial arch 3

Branchial groove 2

Hyoid (branchial arch 2)

Branchial groove 1

Mandibular (branchial arch 1)

Maxillary process

Infundibulum

Facial V.

Aortic arch 4

Closing plate

Pharynx

Aortic arch 3

Ant. vitelline V.

Aortic arch 2

Thyreoglossal duct

Aortic arch 1

Mouth

Ectoderm

Int. carotid A.

Ant. cerebral A.

Diencephalon

Mesoderm

Diocoele

Ant. cardinal V. cephalic portion

Marginal layer

Section 12

Development of the chick embryo VI
x 40 x. sec.

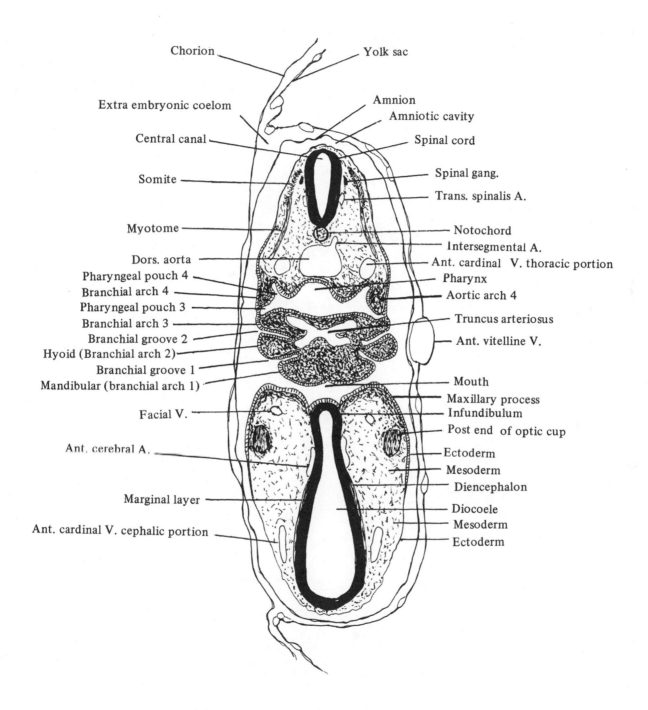

Chorion

Yolk sac

Extra embryonic coelom

Amnion
Amniotic cavity

Central canal

Spinal cord

Somite

Spinal gang.

Trans. spinalis A.

Myotome

Notochord
Intersegmental A.

Dors. aorta

Ant. cardinal V. thoracic portion

Pharyngeal pouch 4

Pharynx

Branchial arch 4

Aortic arch 4

Pharyngeal pouch 3

Branchial arch 3

Truncus arteriosus

Branchial groove 2

Ant. vitelline V.

Hyoid (Branchial arch 2)

Branchial groove 1

Mandibular (branchial arch 1)

Mouth

Maxillary process
Infundibulum
Post end of optic cup

Facial V.

Ant. cerebral A.

Ectoderm
Mesoderm
Diencephalon

Marginal layer

Diocoele
Mesoderm
Ectoderm

Ant. cardinal V. cephalic portion

Section 13

Development of the chick embryo VI
x 40 x. sec.

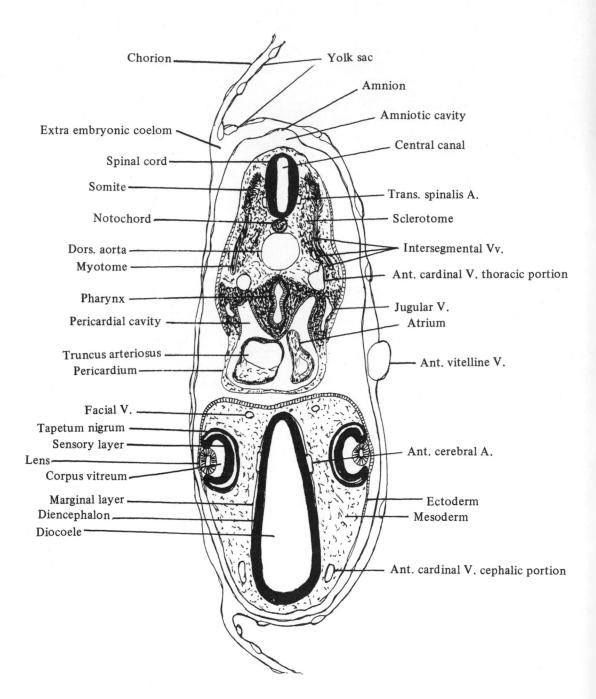

Chorion

Yolk sac

Amnion

Amniotic cavity

Extra embryonic coelom

Central canal

Spinal cord

Somite

Trans. spinalis A.

Notochord

Sclerotome

Dors. aorta

Intersegmental Vv.

Myotome

Ant. cardinal V. thoracic portion

Pharynx

Pericardial cavity

Jugular V.

Atrium

Truncus arteriosus

Pericardium

Ant. vitelline V.

Facial V.

Tapetum nigrum

Sensory layer

Lens

Ant. cerebral A.

Corpus vitreum

Marginal layer

Diencephalon

Ectoderm

Mesoderm

Diocoele

Ant. cardinal V. cephalic portion

Section 14

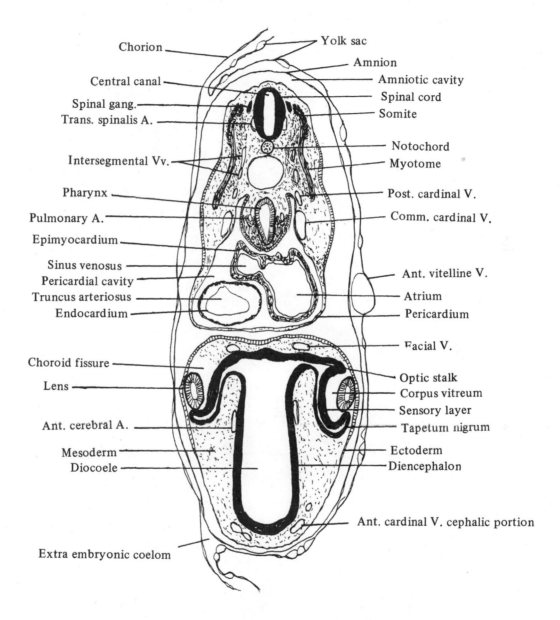

Chorion

Yolk sac

Central canal

Amnion

Spinal gang.

Amniotic cavity

Trans. spinalis A.

Spinal cord

Somite

Intersegmental Vv.

Notochord

Myotome

Pharynx

Post. cardinal V.

Pulmonary A.

Comm. cardinal V.

Epimyocardium

Sinus venosus

Pericardial cavity

Ant. vitelline V.

Truncus arteriosus

Atrium

Endocardium

Pericardium

Facial V.

Choroid fissure

Lens

Optic stalk

Corpus vitreum

Sensory layer

Ant. cerebral A.

Tapetum nigrum

Mesoderm

Ectoderm

Diocoele

Diencephalon

Ant. cardinal V. cephalic portion

Extra embryonic coelom

Section 15

Development of the chick embryo VI
x 40 x. sec.

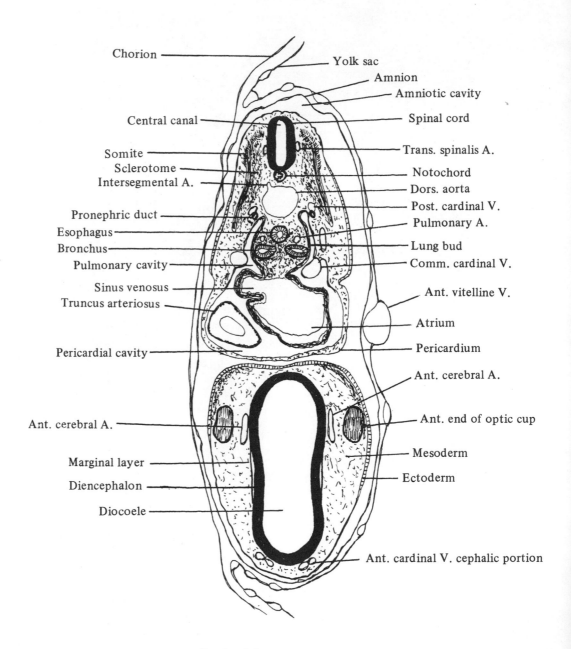

Chorion —————————

Yolk sac

Amnion

Amniotic cavity

Central canal ——————

Spinal cord

Trans. spinalis A.

Somite ——————

Notochord

Sclerotome ————

Dors. aorta

Intersegmental A. ————

Post. cardinal V.

Pronephric duct ————

Pulmonary A.

Esophagus ——————

Lung bud

Bronchus ——————

Comm. cardinal V.

Pulmonary cavity ————

Sinus venosus ————

Ant. vitelline V.

Truncus arteriosus ————

Atrium

Pericardium

Pericardial cavity ————

Ant. cerebral A.

Ant. cerebral A. ————

Ant. end of optic cup

Marginal layer ————

Mesoderm

Diencephalon ————

Ectoderm

Diocoele ————

Ant. cardinal V. cephalic portion

Section 16

Development of the chick embryo VI
x 40 x. sec.

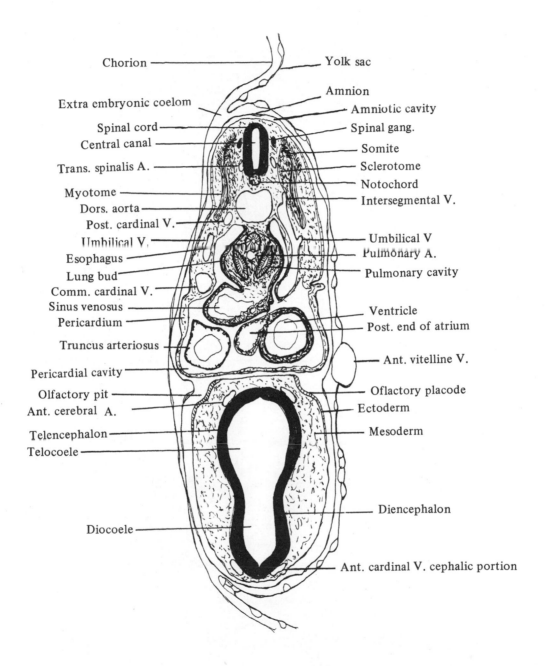

Chorion ——————————— Yolk sac

Amnion

Extra embryonic coelom ————— Amniotic cavity

Spinal cord —————— Spinal gang.
Central canal —————— Somite
Trans. spinalis A. —————— Sclerotome
Notochord
Myotome —————— Intersegmental V.
Dors. aorta ——————
Post. cardinal V. ——————
Umbilical V. —————— Umbilical V
Esophagus —————— Pulmonary A.
Lung bud —————— Pulmonary cavity
Comm. cardinal V. ——————
Sinus venosus ——————
Pericardium —————— Ventricle
Post. end of atrium
Truncus arteriosus ——————
Ant. vitelline V.
Pericardial cavity ——————
Olfactory pit —————— Oflactory placode
Ant. cerebral A. —————— Ectoderm
Telencephalon —————— Mesoderm
Telocoele ——————

Diencephalon

Diocoele ——————

Ant. cardinal V. cephalic portion

Section 17

Development of the chick embryo VI
x 40 x. sec.

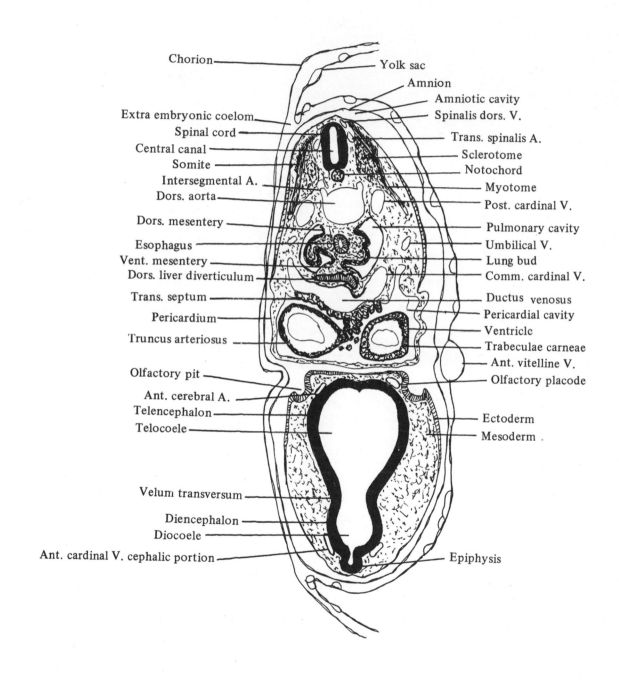

Chorion — Yolk sac
Amnion
Amniotic cavity
Extra embryonic coelom — Spinalis dors. V.
Spinal cord — Trans. spinalis A.
Central canal — Sclerotome
Somite — Notochord
Intersegmental A. — Myotome
Dors. aorta — Post. cardinal V.
Dors. mesentery — Pulmonary cavity
Esophagus — Umbilical V.
Vent. mesentery — Lung bud
Dors. liver diverticulum — Comm. cardinal V.
Trans. septum — Ductus venosus
Pericardium — Pericardial cavity
Truncus arteriosus — Ventricle
Trabeculae carneae
Ant. vitelline V.
Olfactory pit — Olfactory placode
Ant. cerebral A.
Telencephalon — Ectoderm
Telocoele — Mesoderm
Velum transversum
Diencephalon
Diocoele
Ant. cardinal V. cephalic portion — Epiphysis

Section 18

Development of the chick embryo VI
x 40 x. sec.

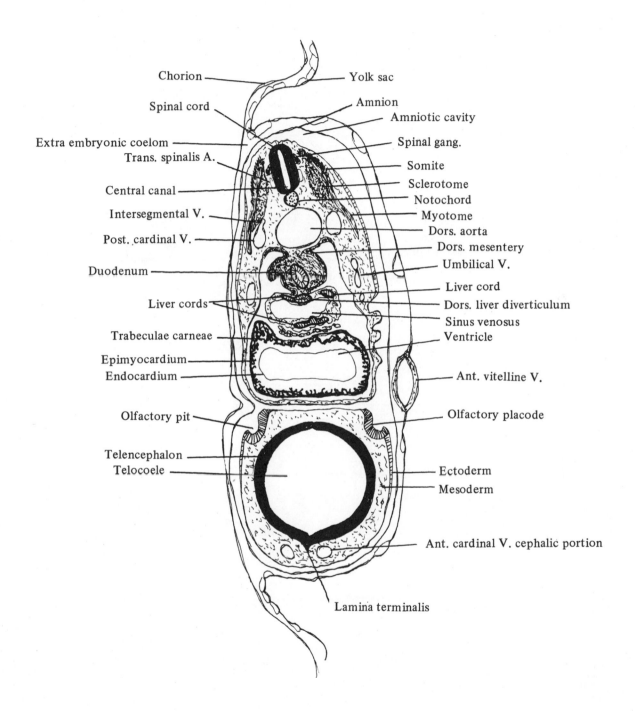

Chorion — Yolk sac
Spinal cord — Amnion
— Amniotic cavity
Extra embryonic coelom — Spinal gang.
Trans. spinalis A. — Somite
— Sclerotome
Central canal — Notochord
Intersegmental V. — Myotome
Post. cardinal V. — Dors. aorta
— Dors. mesentery
Duodenum — Umbilical V.
— Liver cord
Liver cords — Dors. liver diverticulum
— Sinus venosus
Trabeculae carneae — Ventricle
Epimyocardium
Endocardium — Ant. vitelline V.
Olfactory pit — Olfactory placode
Telencephalon — Ectoderm
Telocoele — Mesoderm
— Ant. cardinal V. cephalic portion
Lamina terminalis

Section 19

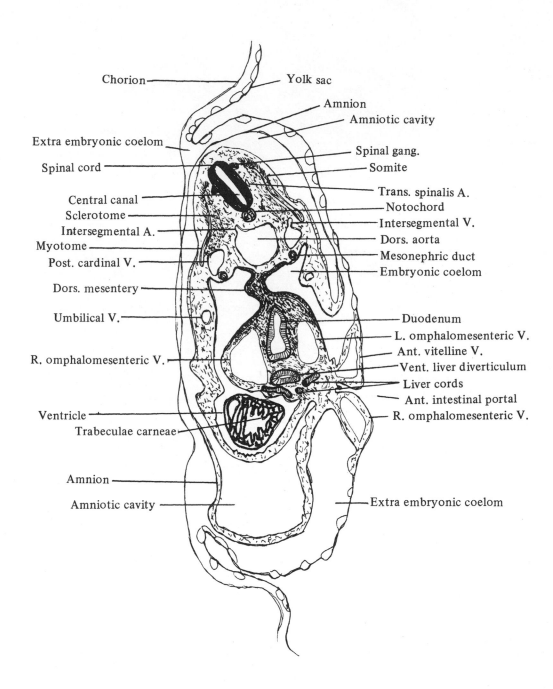

Chorion — Yolk sac

Amnion
Amniotic cavity

Extra embryonic coelom

Spinal gang.

Spinal cord — Somite

Trans. spinalis A.

Central canal — Notochord

Sclerotome — Intersegmental V.

Intersegmental A. — Dors. aorta

Myotome — Mesonephric duct

Post. cardinal V. — Embryonic coelom

Dors. mesentery

Umbilical V. — Duodenum

L. omphalomesenteric V.

Ant. vitelline V.

R. omphalomesenteric V. — Vent. liver diverticulum

Liver cords

Ant. intestinal portal

Ventricle — R. omphalomesenteric V.

Trabeculae carneae

Amnion

Amniotic cavity — Extra embryonic coelom

Section 20

190

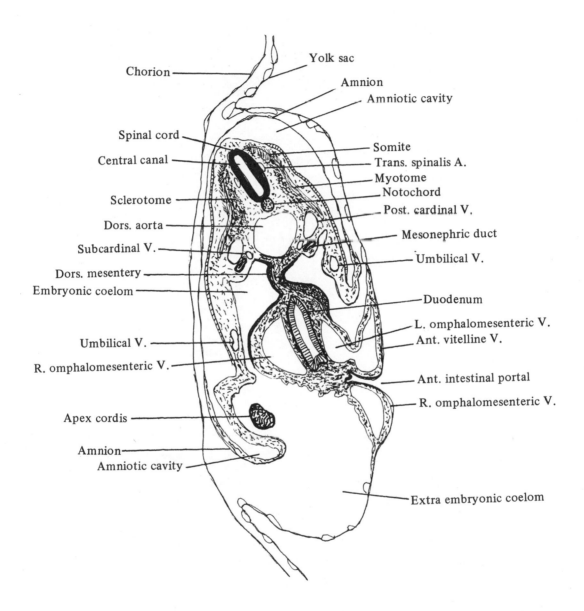

Chorion

Yolk sac

Amnion

Amniotic cavity

Spinal cord

Central canal

Somite

Trans. spinalis A.

Myotome

Notochord

Sclerotome

Post. cardinal V.

Dors. aorta

Mesonephric duct

Subcardinal V.

Umbilical V.

Dors. mesentery

Embryonic coelom

Duodenum

L. omphalomesenteric V.

Ant. vitelline V.

Umbilical V.

R. omphalomesenteric V.

Ant. intestinal portal

R. omphalomesenteric V.

Apex cordis

Amnion

Amniotic cavity

Extra embryonic coelom

Section 21

Development of the chick embryo VI
x 40 x. sec.

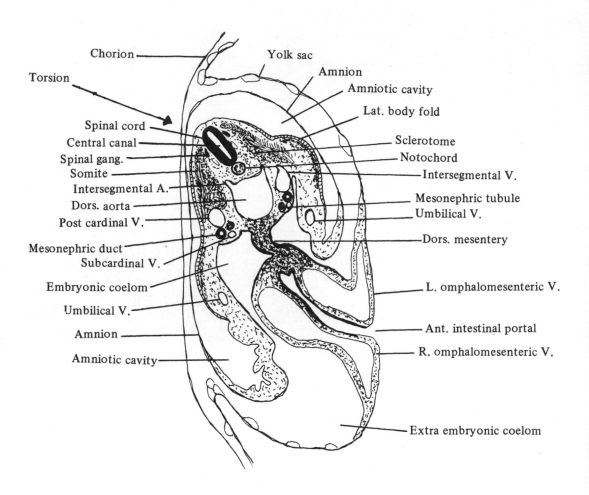

Chorion

Torsion

Yolk sac

Amnion

Amniotic cavity

Lat. body fold

Spinal cord

Central canal

Spinal gang.

Somite

Intersegmental A.

Dors. aorta

Post cardinal V.

Mesonephric duct

Subcardinal V.

Embryonic coelom

Umbilical V.

Amnion

Amniotic cavity

Sclerotome

Notochord

Intersegmental V.

Mesonephric tubule

Umbilical V.

Dors. mesentery

L. omphalomesenteric V.

Ant. intestinal portal

R. omphalomesenteric V.

Extra embryonic coelom

Section 22

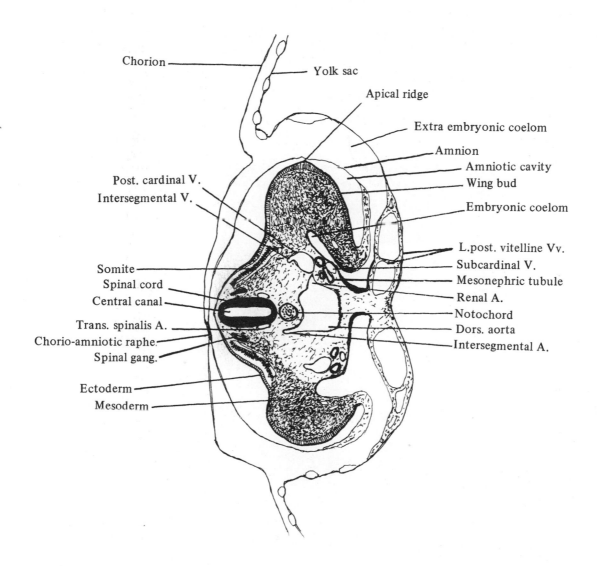

Chorion
Yolk sac
Apical ridge
Extra embryonic coelom
Amnion
Amniotic cavity
Wing bud
Embryonic coelom
Post. cardinal V.
Intersegmental V.
L.post. vitelline Vv.
Subcardinal V.
Mesonephric tubule
Somite
Spinal cord
Renal A.
Central canal
Notochord
Trans. spinalis A.
Dors. aorta
Chorio-amniotic raphe.
Intersegmental A.
Spinal gang.
Ectoderm
Mesoderm

Section 23

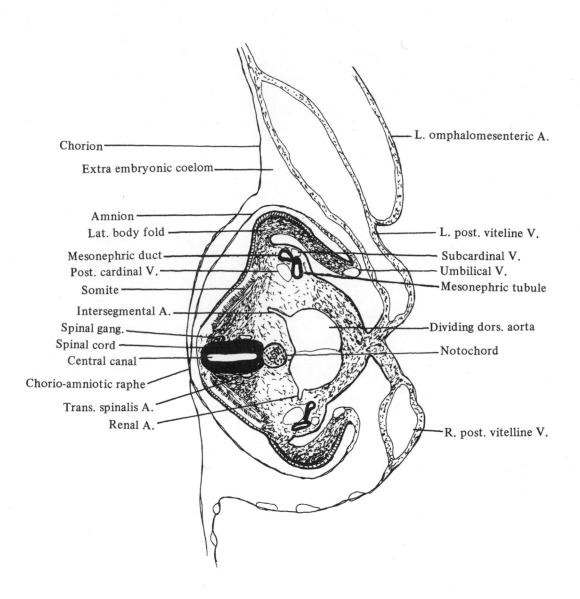

Chorion
Extra embryonic coelom
L. omphalomesenteric A.

Amnion
Lat. body fold
L. post. viteline V.

Mesonephric duct
Post. cardinal V.
Subcardinal V.
Umbilical V.
Mesonephric tubule

Somite
Intersegmental A.
Spinal gang.
Dividing dors. aorta
Spinal cord
Notochord
Central canal
Chorio-amniotic raphe
Trans. spinalis A.
Renal A.
R. post. vitelline V.

Section 24

Development of the chick embryo VI
x 40 x. sec.

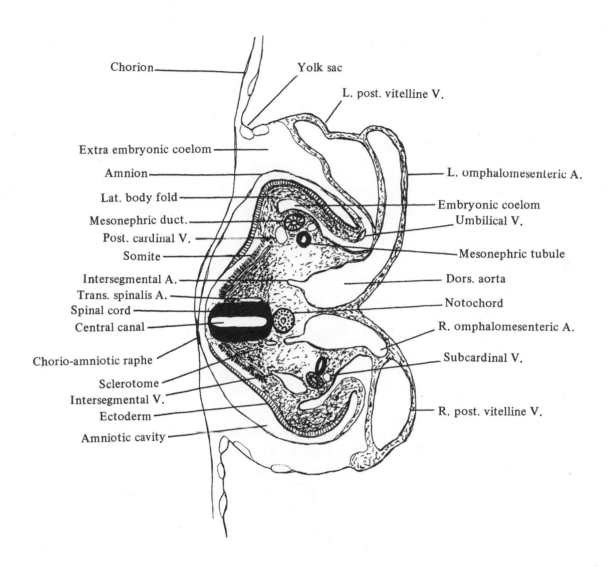

Chorion

Yolk sac

L. post. vitelline V.

Extra embryonic coelom

L. omphalomesenteric A.

Amnion

Lat. body fold

Embryonic coelom

Mesonephric duct.

Umbilical V.

Post. cardinal V.

Somite

Mesonephric tubule

Intersegmental A.

Dors. aorta

Trans. spinalis A.

Notochord

Spinal cord

Central canal

R. omphalomesenteric A.

Chorio-amniotic raphe

Subcardinal V.

Sclerotome

Intersegmental V.

Ectoderm

R. post. vitelline V.

Amniotic cavity

Section 25

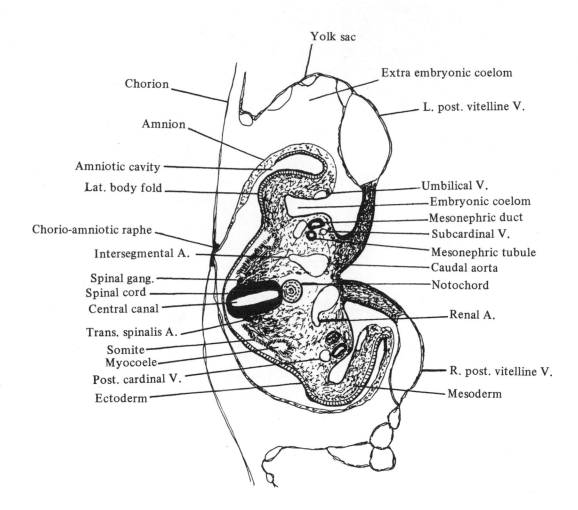

Yolk sac

Chorion

Extra embryonic coelom

L. post. vitelline V.

Amnion

Amniotic cavity

Lat. body fold

Umbilical V.

Embryonic coelom

Mesonephric duct

Chorio-amniotic raphe

Subcardinal V.

Intersegmental A.

Mesonephric tubule

Caudal aorta

Spinal gang.

Notochord

Spinal cord

Central canal

Renal A.

Trans. spinalis A.

Somite

Myocoele

R. post. vitelline V.

Post. cardinal V.

Mesoderm

Ectoderm

Section 26

Development of the chick embryo VI
x 40 x. sec.

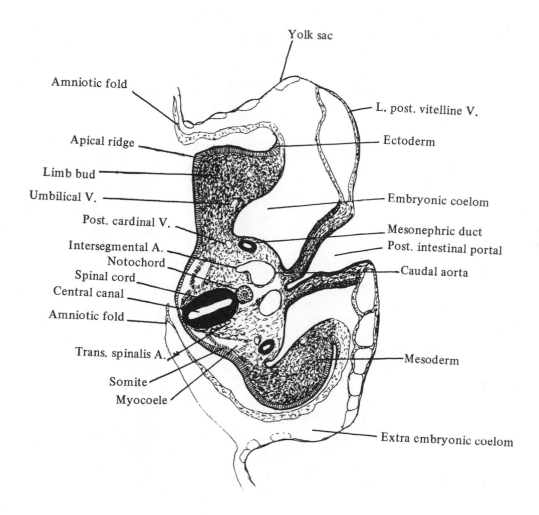

Yolk sac

Amniotic fold

Apical ridge

Limb bud

Umbilical V.

Post. cardinal V.

Intersegmental A.

Notochord

Spinal cord

Central canal

Amniotic fold

Trans. spinalis A.

Somite

Myocoele

L. post. vitelline V.

Ectoderm

Embryonic coelom

Mesonephric duct

Post. intestinal portal

Caudal aorta

Mesoderm

Extra embryonic coelom

Section 27

Development of the chick embryo VI
x 40 x. sec.

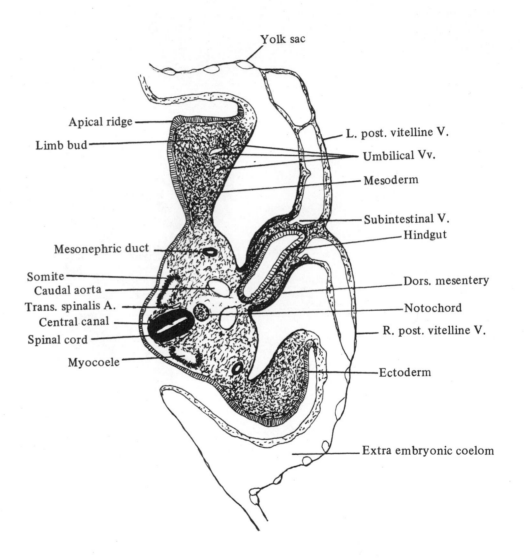

Yolk sac

Apical ridge

Limb bud

L. post. vitelline V.

Umbilical Vv.

Mesoderm

Subintestinal V.

Hindgut

Mesonephric duct

Somite

Dors. mesentery

Caudal aorta

Trans. spinalis A.

Notochord

Central canal

Spinal cord

R. post. vitelline V.

Myocoele

Ectoderm

Extra embryonic coelom

Section 28

Development of the chick embryo VI
x 40 x. sec.

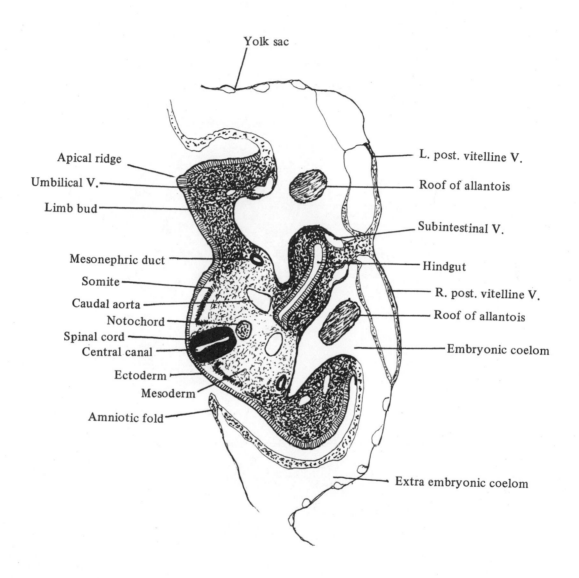

Yolk sac

Apical ridge
Umbilical V.
Limb bud

Mesonephric duct
Somite
Caudal aorta
Notochord
Spinal cord
Central canal
Ectoderm
Mesoderm
Amniotic fold

L. post. vitelline V.
Roof of allantois
Subintestinal V.
Hindgut
R. post. vitelline V.
Roof of allantois
Embryonic coelom

Extra embryonic coelom

Section 29

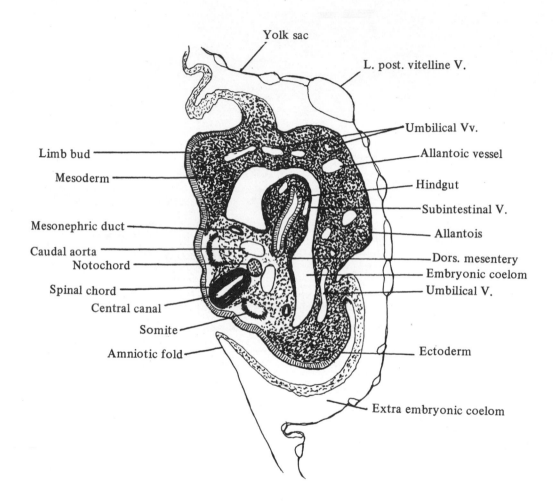

Yolk sac

L. post. vitelline V.

Umbilical Vv.

Limb bud

Allantoic vessel

Mesoderm

Hindgut

Subintestinal V.

Mesonephric duct

Allantois

Caudal aorta

Dors. mesentery

Notochord

Embryonic coelom

Spinal chord

Umbilical V.

Central canal

Somite

Amniotic fold

Ectoderm

Extra embryonic coelom

Section 30

Development of the chick embryo VI
x 40 x. sec.

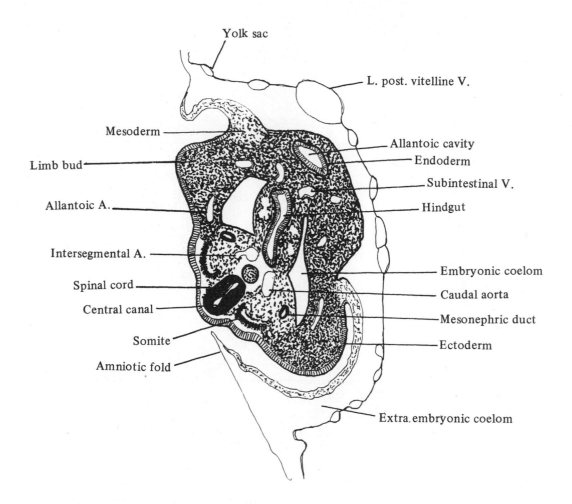

Section 31

Development of the chick embryo VI
x 40 x. sec.

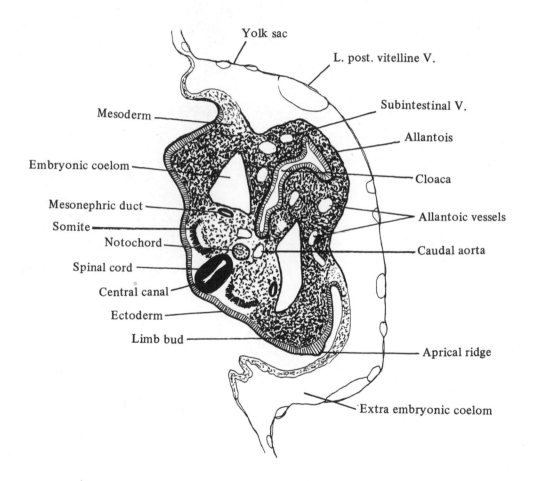

Yolk sac

L. post. vitelline V.

Mesoderm

Subintestinal V.

Allantois

Embryonic coelom

Cloaca

Mesonephric duct

Somite

Allantoic vessels

Notochord

Caudal aorta

Spinal cord

Central canal

Ectoderm

Limb bud

Aprical ridge

Extra embryonic coelom

Section 32

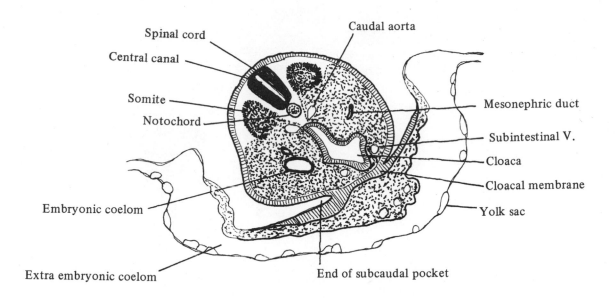

Spinal cord

Central canal

Somite

Notochord

Caudal aorta

Mesonephric duct

Subintestinal V.

Cloaca

Cloacal membrane

Yolk sac

Embryonic coelom

Extra embryonic coelom

End of subcaudal pocket

Section 33

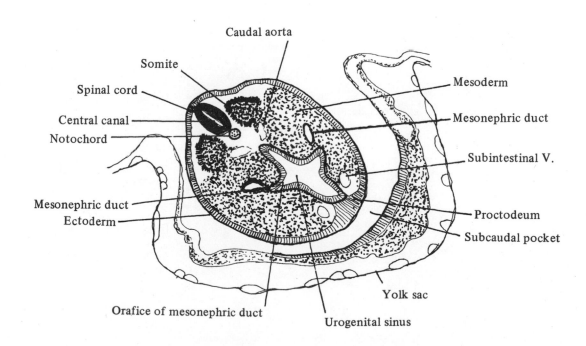

Caudal aorta

Somite

Spinal cord

Central canal

Notochord

Mesoderm

Mesonephric duct

Subintestinal V.

Mesonephric duct

Ectoderm

Proctodeum

Subcaudal pocket

Orafice of mesonephric duct

Urogenital sinus

Yolk sac

Section 34

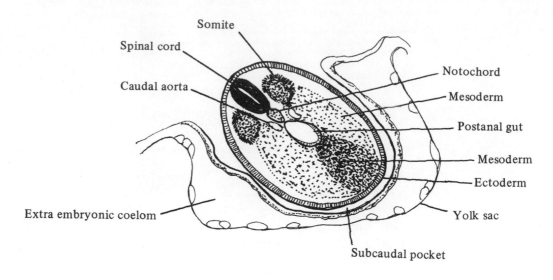

Somite

Spinal cord

Caudal aorta

Notochord

Mesoderm

Postanal gut

Mesoderm

Ectoderm

Extra embryonic coelom

Yolk sac

Subcaudal pocket

Section 35

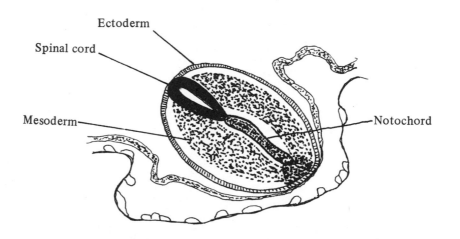

Ectoderm

Spinal cord

Mesoderm

Notochord

Section 36

Development of the Chick Embryo VII
96-Hour Chick Embryo

Development of the chick embryo VII
x 20 w.m.

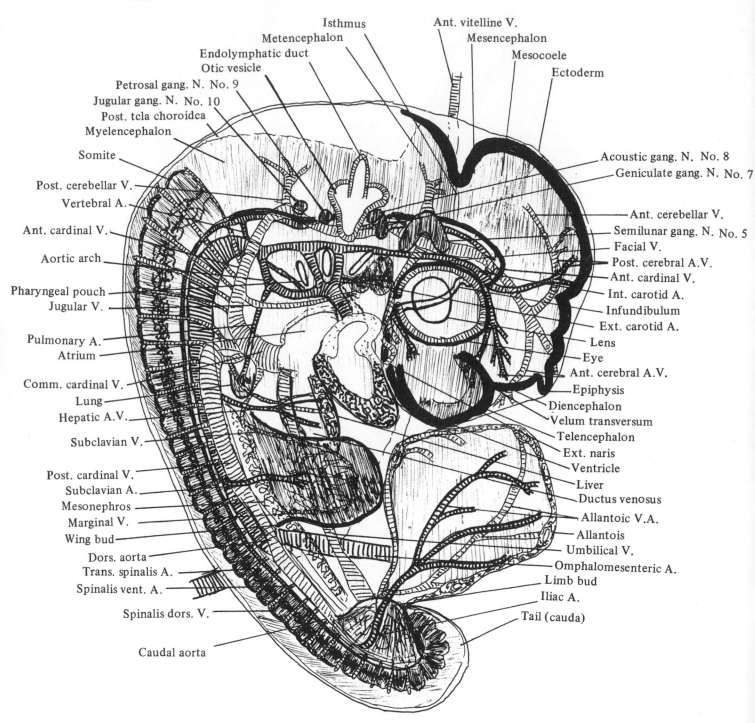

Isthmus
Metencephalon
Endolymphatic duct
Otic vesicle
Petrosal gang. N. No. 9
Jugular gang. N. No. 10
Post. tela choroidea
Myelencephalon
Somite
Post. cerebellar V.
Vertebral A.
Ant. cardinal V.
Aortic arch
Pharyngeal pouch
Jugular V.
Pulmonary A.
Atrium
Comm. cardinal V.
Lung
Hepatic A.V.
Subclavian V.
Post. cardinal V.
Subclavian A.
Mesonephros
Marginal V.
Wing bud
Dors. aorta
Trans. spinalis A.
Spinalis vent. A.
Spinalis dors. V.
Caudal aorta

Ant. vitelline V.
Mesencephalon
Mesocoele
Ectoderm
Acoustic gang. N. No. 8
Geniculate gang. N. No. 7
Ant. cerebellar V.
Semilunar gang. N. No. 5
Facial V.
Post. cerebral A.V.
Ant. cardinal V.
Int. carotid A.
Infundibulum
Ext. carotid A.
Lens
Eye
Ant. cerebral A.V.
Epiphysis
Diencephalon
Velum transversum
Telencephalon
Ext. naris
Ventricle
Liver
Ductus venosus
Allantoic V.A.
Allantois
Umbilical V.
Omphalomesenteric A.
Limb bud
Iliac A.
Tail (cauda)

Development of the chick embryo VII
x 20 sag. sec.

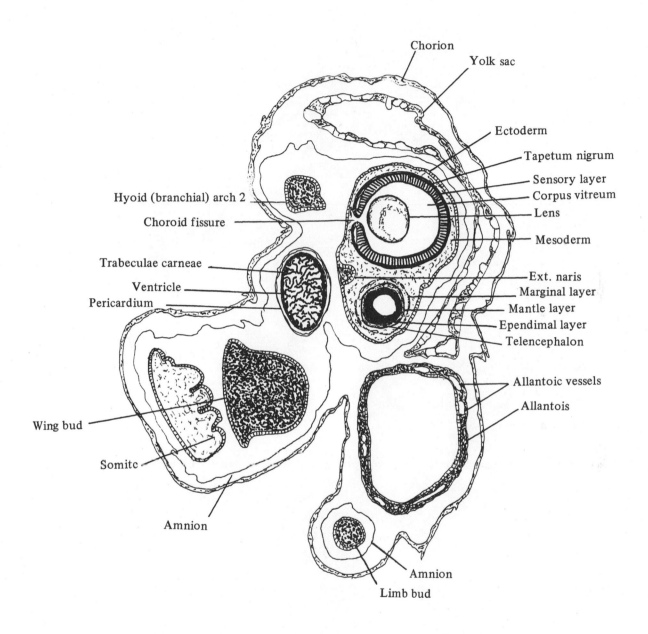

Chorion
Yolk sac
Ectoderm
Tapetum nigrum
Sensory layer
Corpus vitreum
Lens
Mesoderm
Hyoid (branchial) arch 2
Choroid fissure
Ext. naris
Marginal layer
Mantle layer
Ependimal layer
Telencephalon
Trabeculae carneae
Ventricle
Pericardium
Allantoic vessels
Allantois
Wing bud
Somite
Amnion
Amnion
Limb bud

Section A

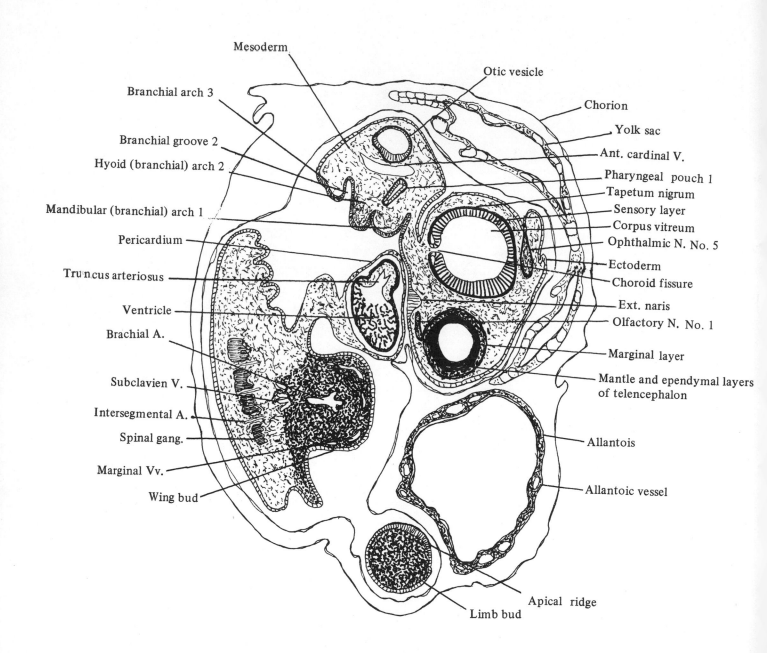

Mesoderm

Branchial arch 3

Branchial groove 2

Hyoid (branchial) arch 2

Mandibular (branchial) arch 1

Pericardium

Truncus arteriosus

Ventricle

Brachial A.

Subclavien V.

Intersegmental A.

Spinal gang.

Marginal Vv.

Wing bud

Otic vesicle

Chorion

Yolk sac

Ant. cardinal V.

Pharyngeal pouch 1

Tapetum nigrum

Sensory layer

Corpus vitreum

Ophthalmic N. No. 5

Ectoderm

Choroid fissure

Ext. naris

Olfactory N. No. 1

Marginal layer

Mantle and ependymal layers
of telencephalon

Allantois

Allantoic vessel

Apical ridge

Limb bud

Section B

Development of the chick embryo VII
x 20 sag. sec.

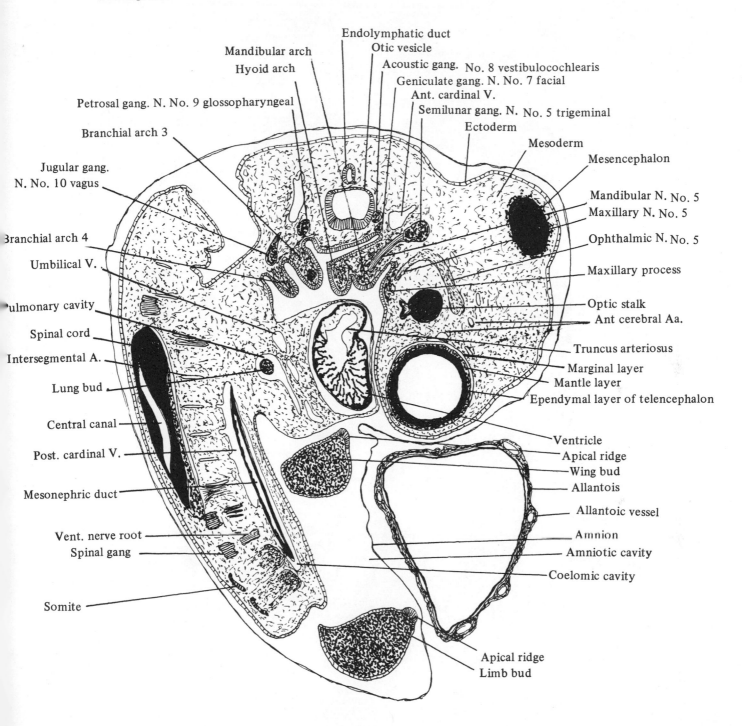

Endolymphatic duct
Otic vesicle
Acoustic gang.
No. 8 vestibulocochlearis
Geniculate gang. N. No. 7 facial
Ant. cardinal V.
Semilunar gang. N. No. 5 trigeminal
Ectoderm
Mesoderm
Mesencephalon

Mandibular arch
Hyoid arch

Petrosal gang. N. No. 9 glossopharyngeal

Branchial arch 3

Jugular gang.
N. No. 10 vagus

Branchial arch 4

Umbilical V.

Pulmonary cavity

Spinal cord

Intersegmental A.

Lung bud

Central canal

Post. cardinal V.

Mesonephric duct

Vent. nerve root
Spinal gang

Somite

Mandibular N. No. 5
Maxillary N. No. 5

Ophthalmic N. No. 5

Maxillary process

Optic stalk
Ant cerebral Aa.

Truncus arteriosus

Marginal layer
Mantle layer
Ependymal layer of telencephalon

Ventricle
Apical ridge
Wing bud
Allantois

Allantoic vessel

Amnion
Amniotic cavity

Coelomic cavity

Apical ridge
Limb bud

Section C

Development of the chick embryo VII

x 20 sag. sec.

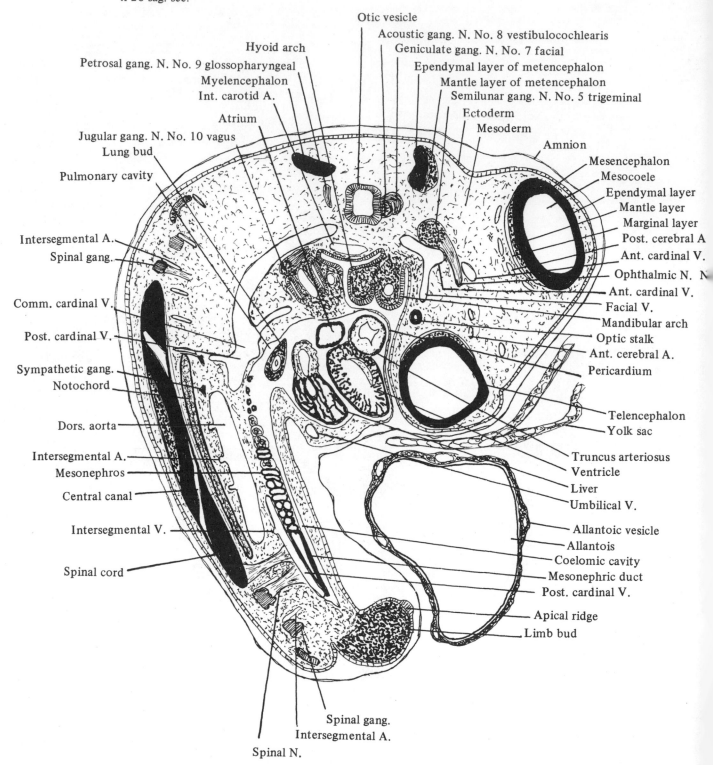

Otic vesicle
Acoustic gang. N. No. 8 vestibulocochlearis
Geniculate gang. N. No. 7 facial
Ependymal layer of metencephalon
Mantle layer of metencephalon
Semilunar gang. N. No. 5 trigeminal
Ectoderm
Mesoderm
Amnion

Hyoid arch
Petrosal gang. N. No. 9 glossopharyngeal
Myelencephalon
Int. carotid A.
Atrium
Jugular gang. N. No. 10 vagus
Lung bud
Pulmonary cavity

Mesencephalon
Mesocoele
Ependymal layer
Mantle layer
Marginal layer
Post. cerebral A
Ant. cardinal V.
Ophthalmic N. N
Ant. cardinal V.
Facial V.
Mandibular arch
Optic stalk
Ant. cerebral A.
Pericardium

Intersegmental A.
Spinal gang.
Comm. cardinal V.
Post. cardinal V.
Sympathetic gang.
Notochord
Dors. aorta
Intersegmental A.
Mesonephros
Central canal
Intersegmental V.
Spinal cord

Telencephalon
Yolk sac
Truncus arteriosus
Ventricle
Liver
Umbilical V.
Allantoic vesicle
Allantois
Coelomic cavity
Mesonephric duct
Post. cardinal V.
Apical ridge
Limb bud

Spinal gang.
Intersegmental A.
Spinal N.

Section D

210

Development of the chick embryo VII
x 20 sag. sec.

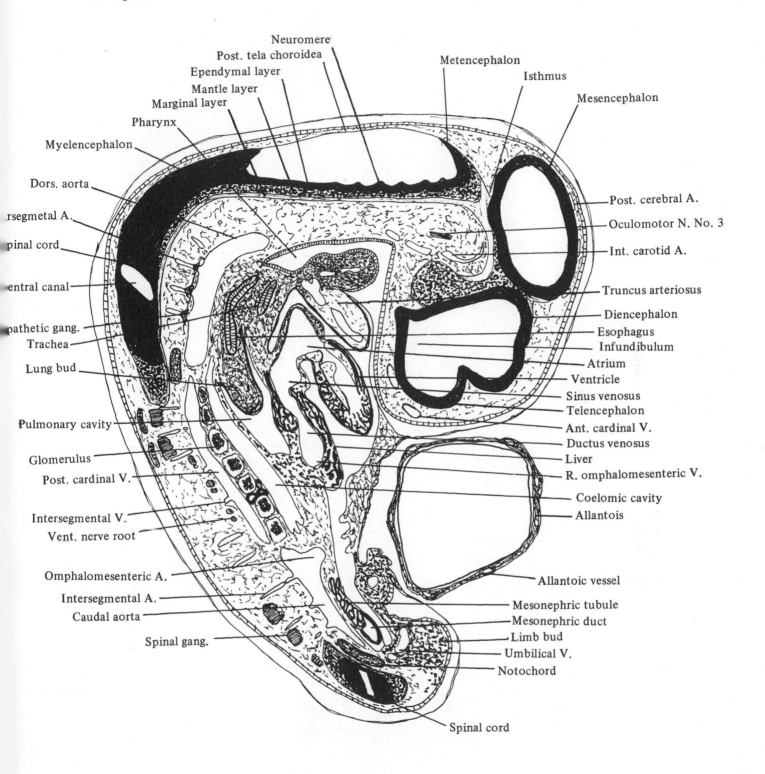

Neuromere
Post. tela choroidea
Ependymal layer
Mantle layer
Marginal layer
Pharynx
Myelencephalon
Dors. aorta
rsegmetal A.
pinal cord
entral canal
pathetic gang.
Trachea
Lung bud
Pulmonary cavity
Glomerulus
Post. cardinal V.
Intersegmental V.
Vent. nerve root
Omphalomesenteric A.
Intersegmental A.
Caudal aorta
Spinal gang.

Metencephalon
Isthmus
Mesencephalon
Post. cerebral A.
Oculomotor N. No. 3
Int. carotid A.
Truncus arteriosus
Diencephalon
Esophagus
Infundibulum
Atrium
Ventricle
Sinus venosus
Telencephalon
Ant. cardinal V.
Ductus venosus
Liver
R. omphalomesenteric V.
Coelomic cavity
Allantois
Allantoic vessel
Mesonephric tubule
Mesonephric duct
Limb bud
Umbilical V.
Notochord

Spinal cord

Section E

Development of the chick embryo VII
x 20 sag. sec.

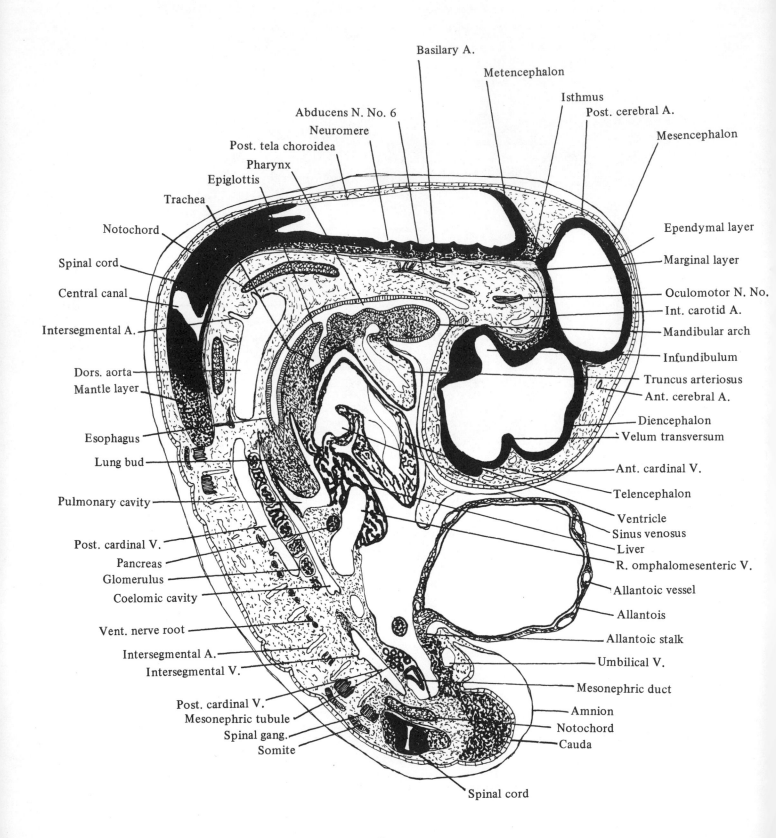

Basilary A.

Metencephalon

Isthmus

Post. cerebral A.

Abducens N. No. 6

Mesencephalon

Neuromere

Post. tela choroidea

Pharynx

Epiglottis

Trachea

Ependymal layer

Notochord

Marginal layer

Spinal cord

Central canal

Oculomotor N. No.

Int. carotid A.

Intersegmental A.

Mandibular arch

Infundibulum

Dors. aorta

Truncus arteriosus

Mantle layer

Ant. cerebral A.

Diencephalon

Esophagus

Velum transversum

Lung bud

Ant. cardinal V.

Telencephalon

Pulmonary cavity

Ventricle

Sinus venosus

Post. cardinal V.

Liver

Pancreas

R. omphalomesenteric V.

Glomerulus

Allantoic vessel

Coelomic cavity

Allantois

Vent. nerve root

Allantoic stalk

Intersegmental A.

Umbilical V.

Intersegmental V.

Mesonephric duct

Post. cardinal V.

Amnion

Mesonephric tubule

Notochord

Spinal gang.

Cauda

Somite

Spinal cord

Section F
212

Development of the chick embryo VII
x 20 sag. sec.

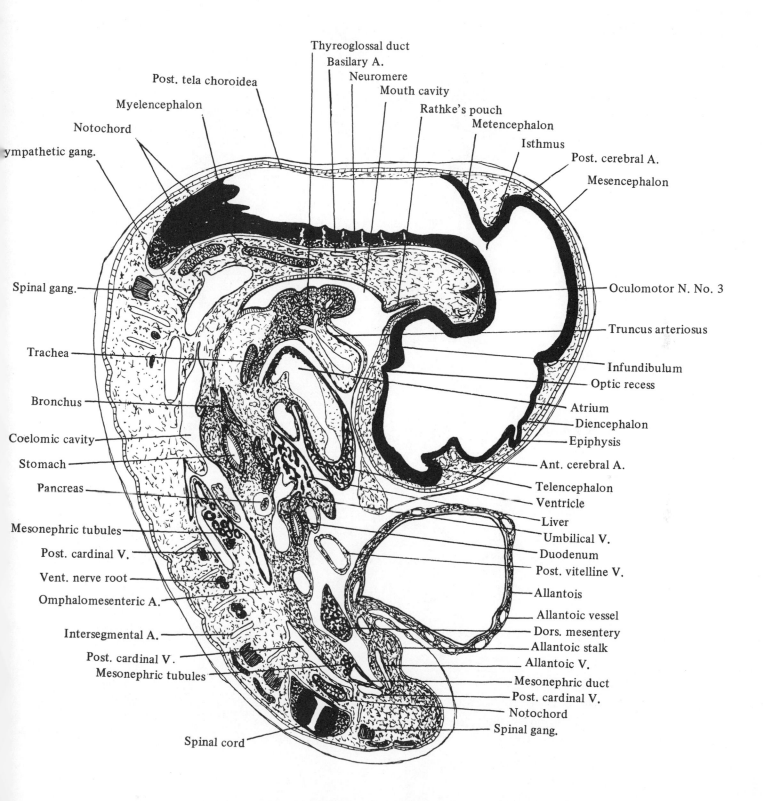

Thyreoglossal duct
Basilary A.
Neuromere
Mouth cavity
Rathke's pouch
Metencephalon
Isthmus
Post. cerebral A.
Mesencephalon

Post. tela choroidea
Myelencephalon
Notochord
ympathetic gang.

Spinal gang.
Trachea
Bronchus
Coelomic cavity
Stomach
Pancreas
Mesonephric tubules
Post. cardinal V.
Vent. nerve root
Omphalomesenteric A.
Intersegmental A.
Post. cardinal V.
Mesonephric tubules
Spinal cord

Oculomotor N. No. 3
Truncus arteriosus
Infundibulum
Optic recess
Atrium
Diencephalon
Epiphysis
Ant. cerebral A.
Telencephalon
Ventricle
Liver
Umbilical V.
Duodenum
Post. vitelline V.
Allantois
Allantoic vessel
Dors. mesentery
Allantoic stalk
Allantoic V.
Mesonephric duct
Post. cardinal V.
Notochord
Spinal gang.

Section G

213

Development of the chick embryo VII
x 20 sag. sec.

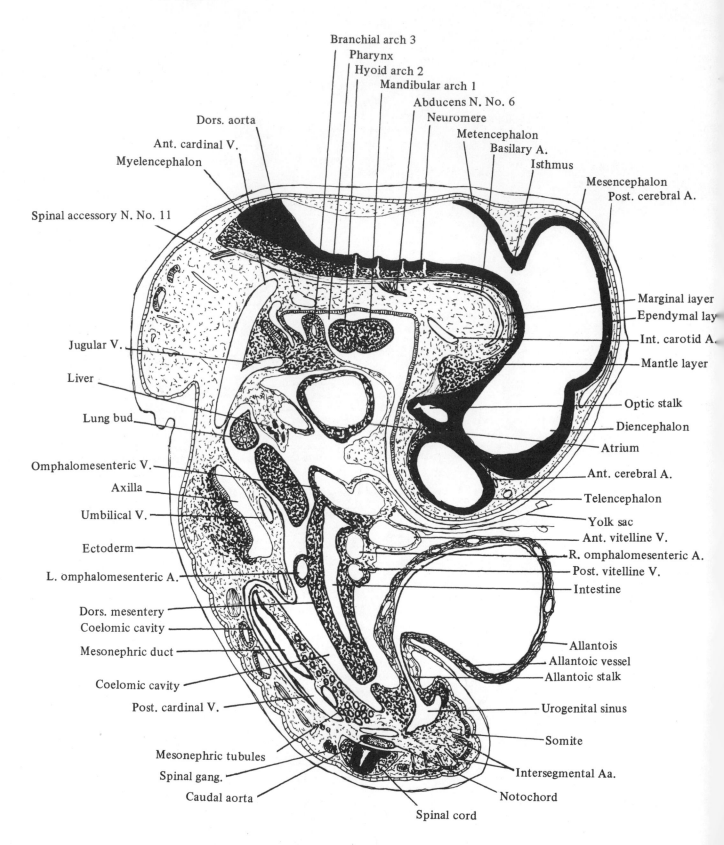

Branchial arch 3
Pharynx
Hyoid arch 2
Mandibular arch 1
Abducens N. No. 6
Neuromere
Metencephalon
Basilary A.
Isthmus
Mesencephalon
Post. cerebral A.

Dors. aorta
Ant. cardinal V.
Myelencephalon

Spinal accessory N. No. 11

Marginal layer
Ependymal lay
Int. carotid A.
Mantle layer

Jugular V.

Liver

Optic stalk
Diencephalon
Atrium
Ant. cerebral A.
Telencephalon
Yolk sac
Ant. vitelline V.
R. omphalomesenteric A.
Post. vitelline V.
Intestine

Lung bud

Omphalomesenteric V.
Axilla
Umbilical V.
Ectoderm
L. omphalomesenteric A.

Dors. mesentery
Coelomic cavity
Mesonephric duct

Coelomic cavity
Post. cardinal V.

Allantois
Allantoic vessel
Allantoic stalk
Urogenital sinus
Somite
Intersegmental Aa.
Notochord

Mesonephric tubules
Spinal gang.
Caudal aorta

Spinal cord

Section H

214

Development of the chick embryo VII
x 20 sag. sec.

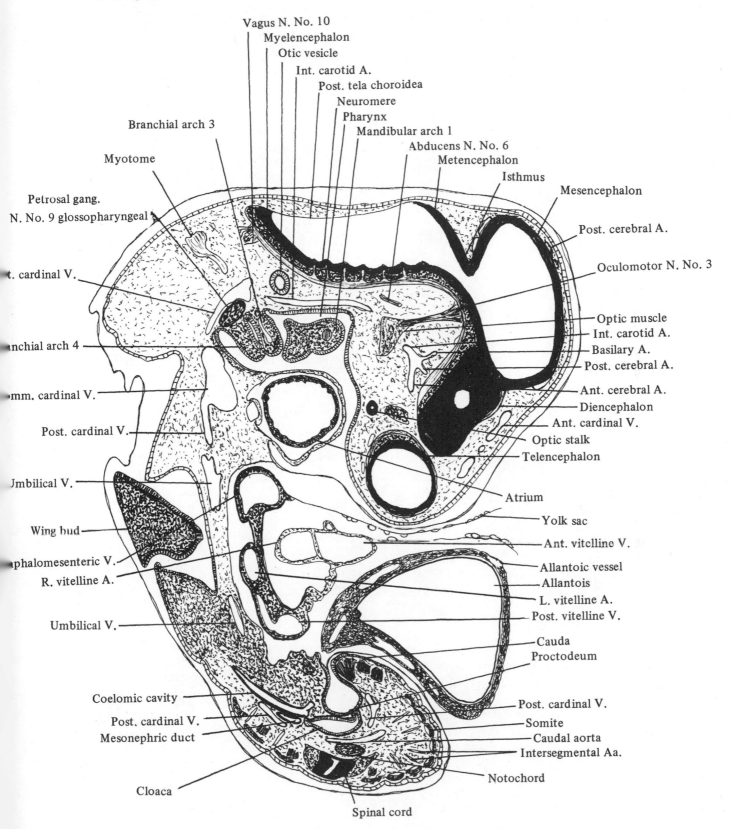

Vagus N. No. 10
Myelencephalon
Otic vesicle
Int. carotid A.
Post. tela choroidea
Neuromere
Pharynx
Mandibular arch 1
Abducens N. No. 6
Metencephalon
Isthmus
Mesencephalon

Branchial arch 3

Myotome

Post. cerebral A.

Petrosal gang.
N. No. 9 glossopharyngeal

Oculomotor N. No. 3

t. cardinal V.

Optic muscle
Int. carotid A.
Basilary A.
Post. cerebral A.

nchial arch 4

Ant. cerebral A.
Diencephalon
Ant. cardinal V.
Optic stalk
Telencephalon

mm. cardinal V.

Post. cardinal V.

Umbilical V.

Atrium

Yolk sac

Wing bud

Ant. vitelline V.

phalomesenteric V.
R. vitelline A.

Allantoic vessel
Allantois
L. vitelline A.
Post. vitelline V.

Umbilical V.

Cauda
Proctodeum

Coelomic cavity

Post. cardinal V.

Post. cardinal V.
Mesonephric duct

Somite
Caudal aorta
Intersegmental Aa.
Notochord

Cloaca

Spinal cord

Section I

Development of the chick embryo VII
x 20 sag. sec.

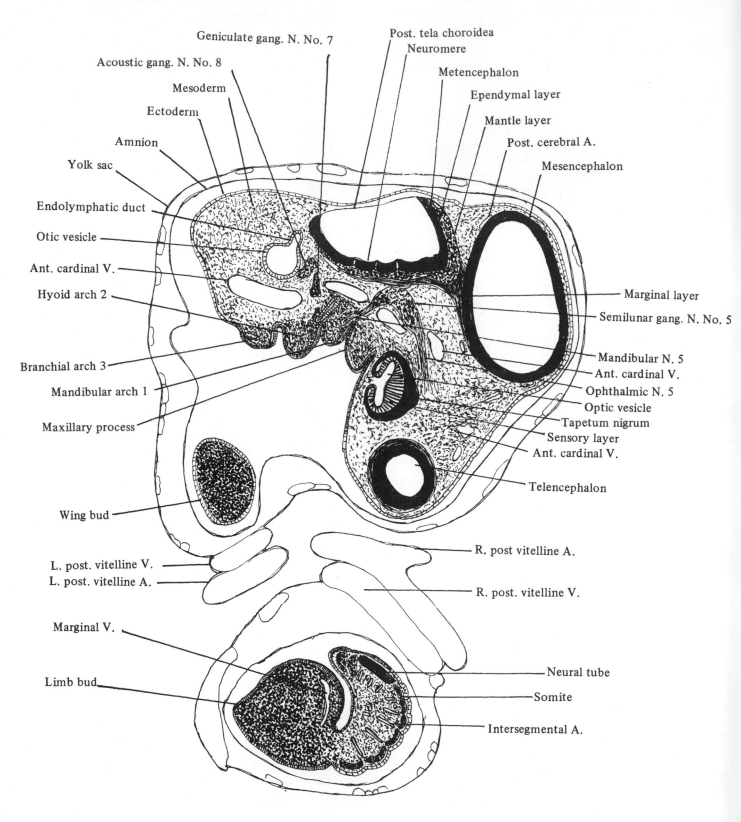

Geniculate gang. N. No. 7

Acoustic gang. N. No. 8

Mesoderm

Ectoderm

Amnion

Yolk sac

Endolymphatic duct

Otic vesicle

Ant. cardinal V.

Hyoid arch 2

Branchial arch 3

Mandibular arch 1

Maxillary process

Wing bud

L. post. vitelline V.

L. post. vitelline A.

Marginal V.

Limb bud

Post. tela choroidea

Neuromere

Metencephalon

Ependymal layer

Mantle layer

Post. cerebral A.

Mesencephalon

Marginal layer

Semilunar gang. N. No. 5

Mandibular N. 5

Ant. cardinal V.

Ophthalmic N. 5

Optic vesicle

Tapetum nigrum

Sensory layer

Ant. cardinal V.

Telencephalon

R. post vitelline A.

R. post. vitelline V.

Neural tube

Somite

Intersegmental A.

Section J

216

Development of the Chick Embryo VIII

High Magnification,
72- and 96-Hour Chick Embryo

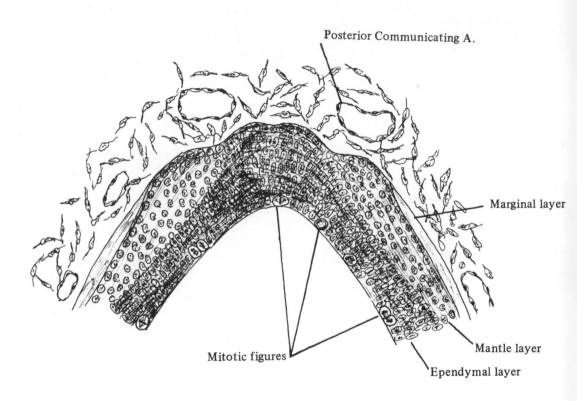

Posterior Communicating A.

Marginal layer

Mantle layer

Ependymal layer

Mitotic figures

Mesencephalon
72-hour chick embryo

Development of the chick embryo VIII
x 400 x. sec.

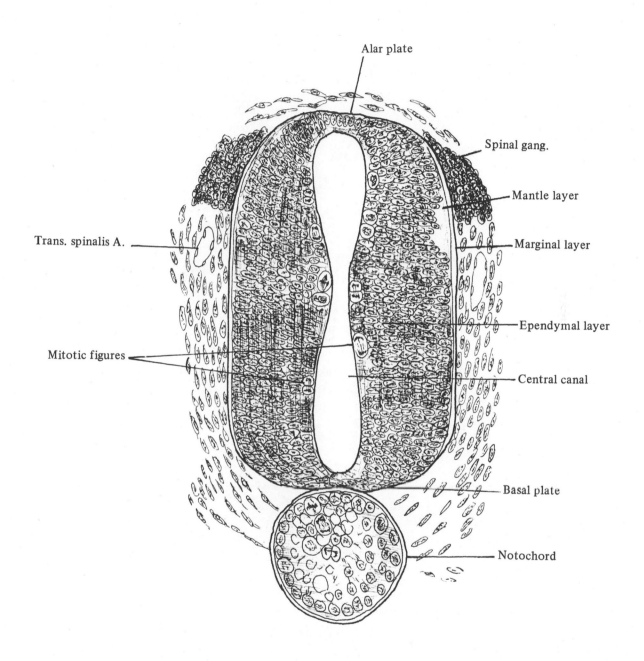

Alar plate

Spinal gang.

Mantle layer

Trans. spinalis A.

Marginal layer

Ependymal layer

Mitotic figures

Central canal

Basal plate

Notochord

Spinal cord

72-hour chick embryo

Development of the chick embryo VIII
x 400 x. sec.

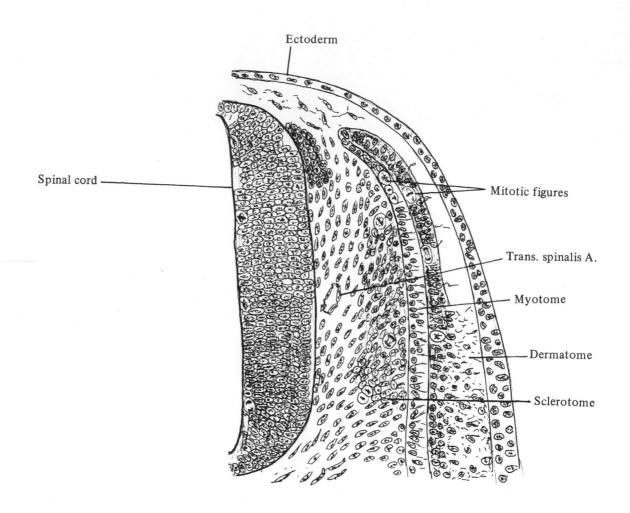

Ectoderm

Spinal cord

Mitotic figures

Trans. spinalis A.

Myotome

Dermatome

Sclerotome

Somite

72-hour chick embryo

Development of the chick embryo VIII
x 400 x. sec.

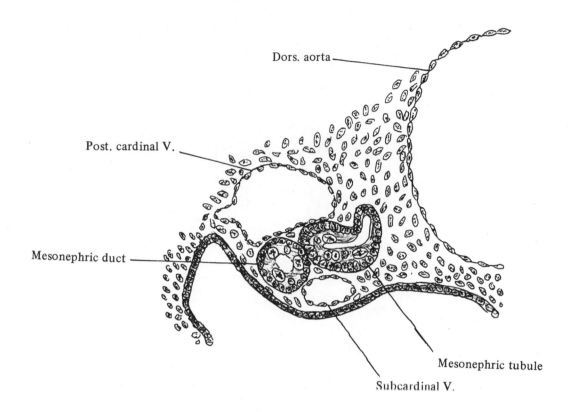

Dors. aorta

Post. cardinal V.

Mesonephric duct

Mesonephric tubule

Subcardinal V.

Mesonephros
72-hour chick embryo

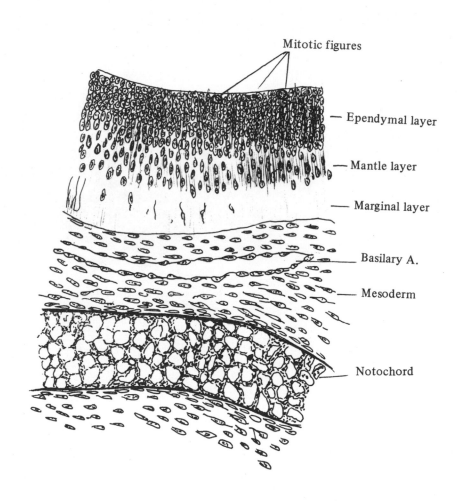

Mitotic figures

Ependymal layer

Mantle layer

Marginal layer

Basilary A.

Mesoderm

Notochord

Myelencephalon

96-hour chick embryo

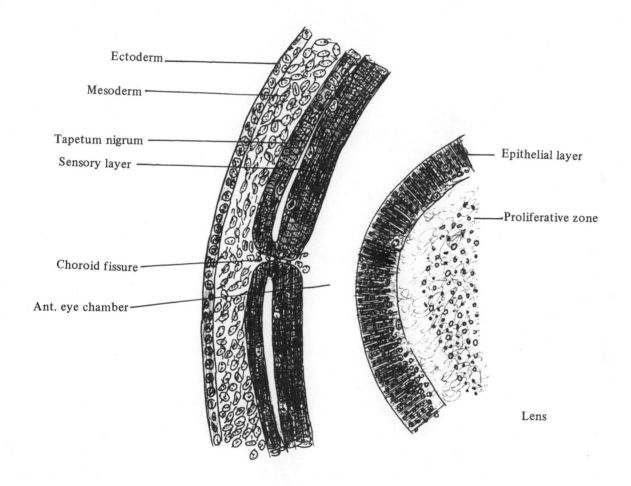

Ectoderm

Mesoderm

Tapetum nigrum

Sensory layer

Choroid fissure

Ant. eye chamber

Epithelial layer

Proliferative zone

Lens

Eye

96-hour chick embryo

Part III

Mammalian Development I

Rat Testis
Cat Ovary
Rat Ovary, Corpus Luteum
Human Ovary, Corpus Luteum
Corpus Albicans
Pregnant Uterus of the Rat

Rat testis
x 40, section

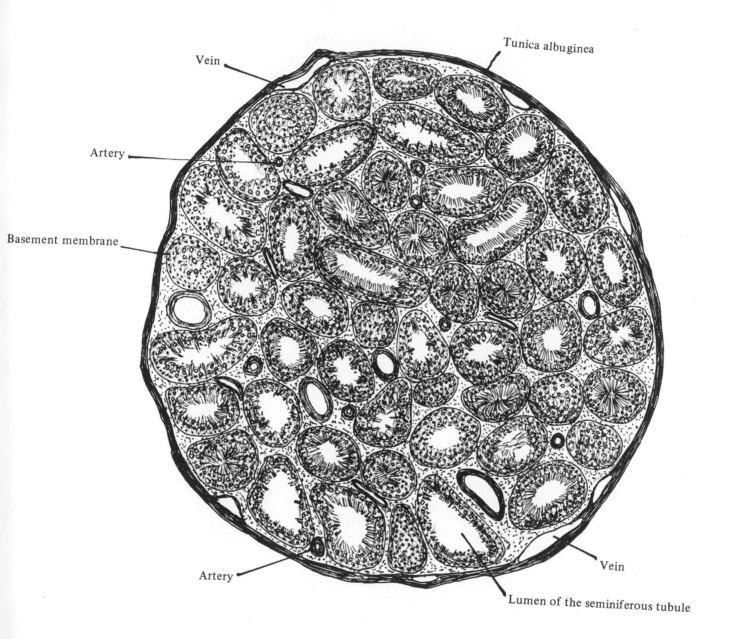

Vein

Tunica albuginea

Artery

Basement membrane

Vein

Artery

Lumen of the seminiferous tubule

Rat testis
x 400, section

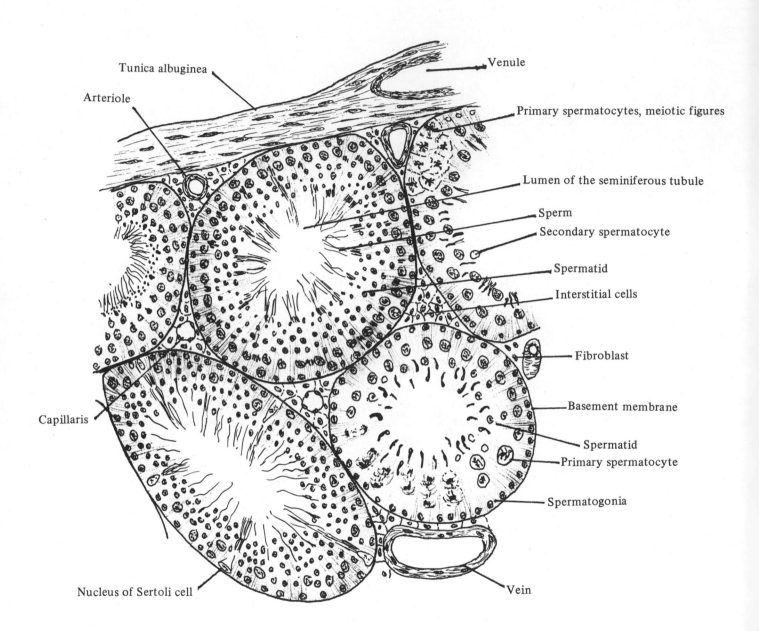

Tunica albuginea

Arteriole

Venule

Primary spermatocytes, meiotic figures

Lumen of the seminiferous tubule

Sperm

Secondary spermatocyte

Spermatid

Interstitial cells

Fibroblast

Basement membrane

Spermatid

Primary spermatocyte

Spermatogonia

Capillaris

Nucleus of Sertoli cell

Vein

Rat testis
x 1,500 section

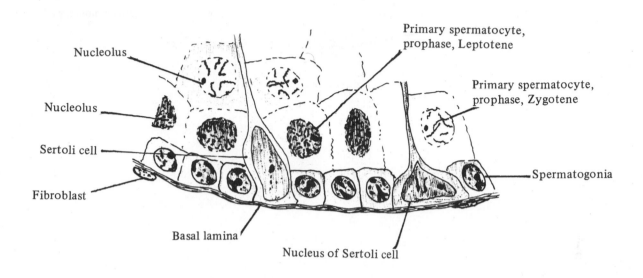

Nucleolus

Nucleolus

Sertoli cell

Fibroblast

Primary spermatocyte,
prophase, Leptotene

Primary spermatocyte,
prophase, Zygotene

Spermatogonia

Basal lamina

Nucleus of Sertoli cell

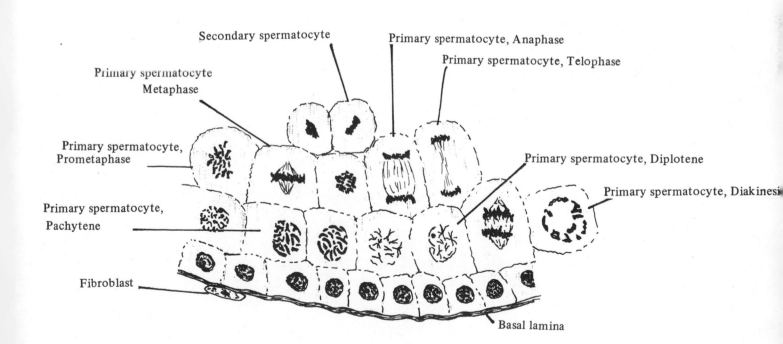

Secondary spermatocyte

Primary spermatocyte, Anaphase

Primary spermatocyte, Telophase

Primary spermatocyte
Metaphase

Primary spermatocyte,
Prometaphase

Primary spermatocyte, Diplotene

Primary spermatocyte, Diakinesis

Primary spermatocyte,
Pachytene

Fibroblast

Basal lamina

229

Rat testis
x 1,500 section

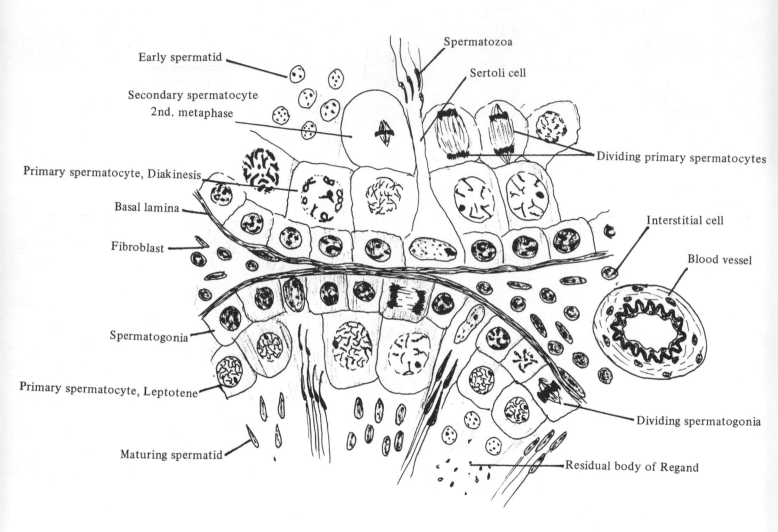

Early spermatid

Secondary spermatocyte
2nd. metaphase

Primary spermatocyte, Diakinesis

Basal lamina

Fibroblast

Spermatogonia

Primary spermatocyte, Leptotene

Maturing spermatid

Spermatozoa

Sertoli cell

Dividing primary spermatocytes

Interstitial cell

Blood vessel

Dividing spermatogonia

Residual body of Regand

Cat ovary
x 40 section

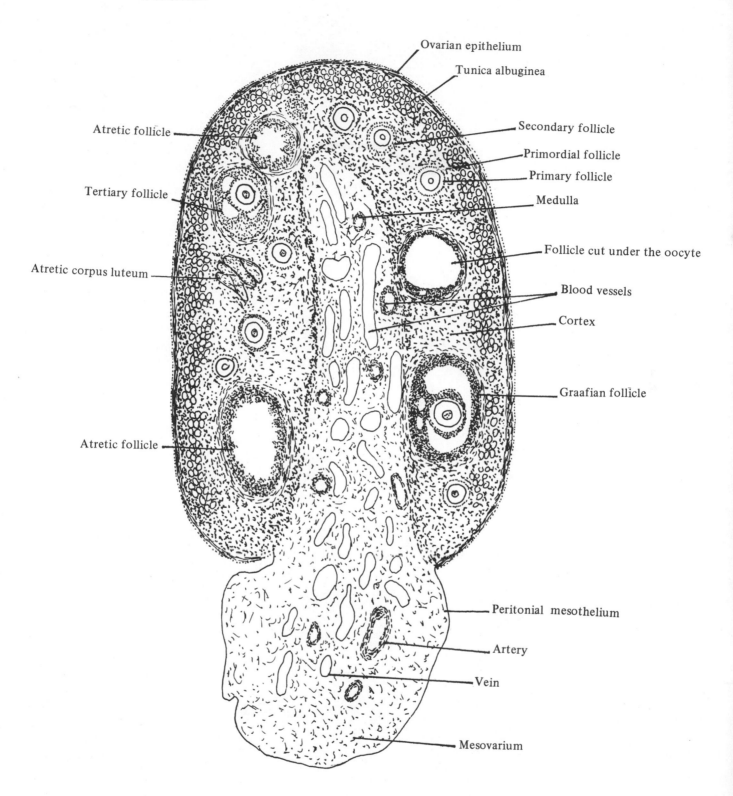

Atretic follicle

Tertiary follicle

Atretic corpus luteum

Atretic follicle

Ovarian epithelium

Tunica albuginea

Secondary follicle

Primordial follicle

Primary follicle

Medulla

Follicle cut under the oocyte

Blood vessels

Cortex

Graafian follicle

Peritonial mesothelium

Artery

Vein

Mesovarium

Cat ovary
x 400 section

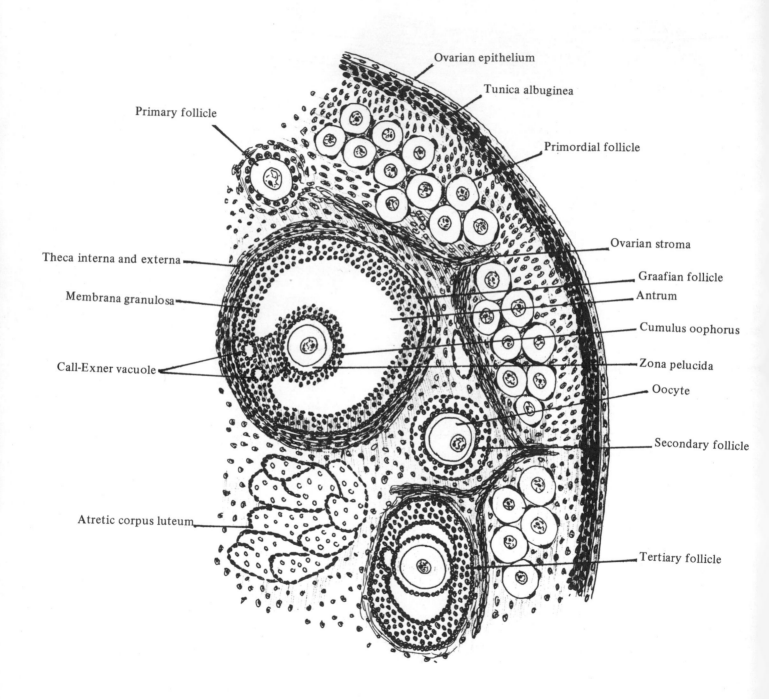

Ovarian epithelium

Tunica albuginea

Primary follicle

Primordial follicle

Theca interna and externa

Ovarian stroma

Graafian follicle

Membrana granulosa

Antrum

Cumulus oophorus

Call-Exner vacuole

Zona pelucida

Oocyte

Secondary follicle

Atretic corpus luteum

Tertiary follicle

Rat ovary, corpus luteum
x 100 section

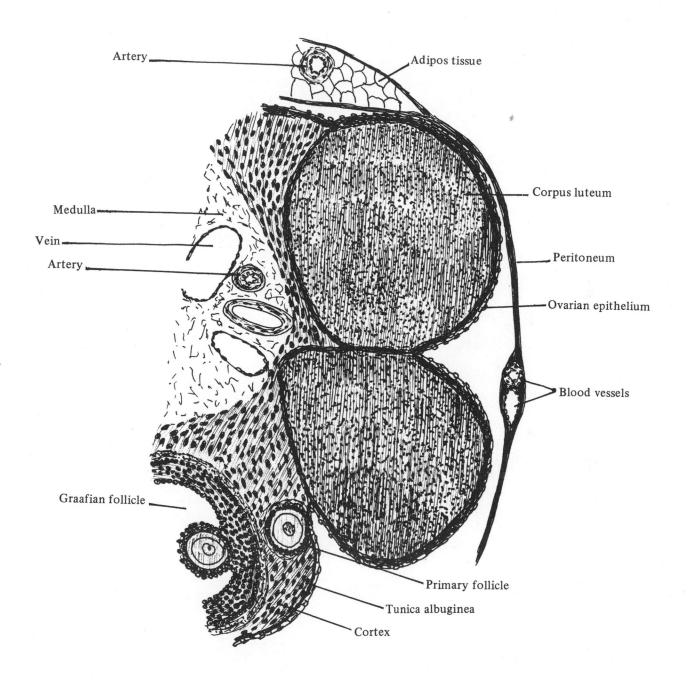

Artery

Adipos tissue

Corpus luteum

Medulla

Peritoneum

Vein

Artery

Ovarian epithelium

Blood vessels

Graafian follicle

Primary follicle

Tunica albuginea

Cortex

Human ovary, corpus luteum,
corpus albicans
x 40 section

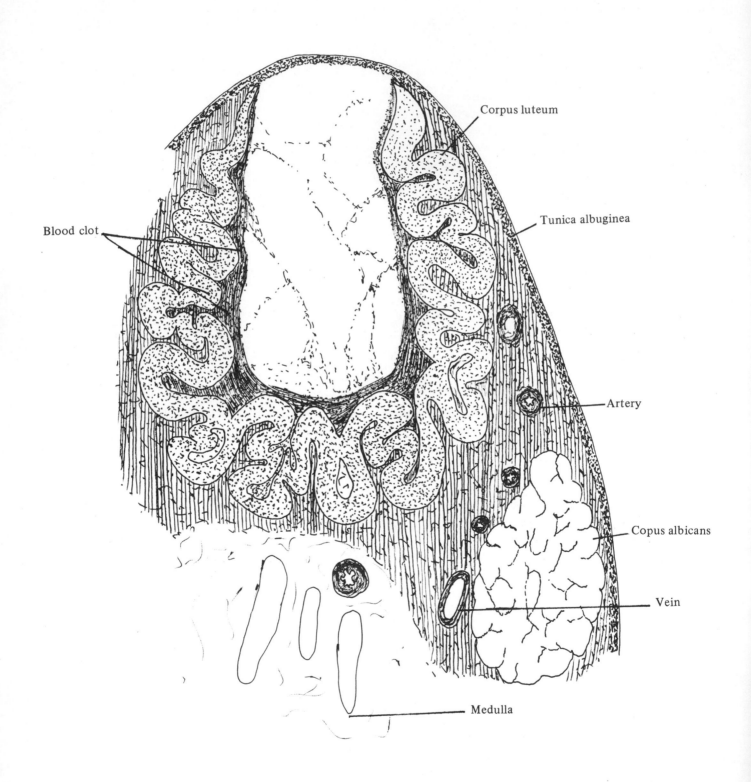

Corpus luteum

Tunica albuginea

Blood clot

Artery

Copus albicans

Vein

Medulla

Pregnant uterus of the rat
x 40, section

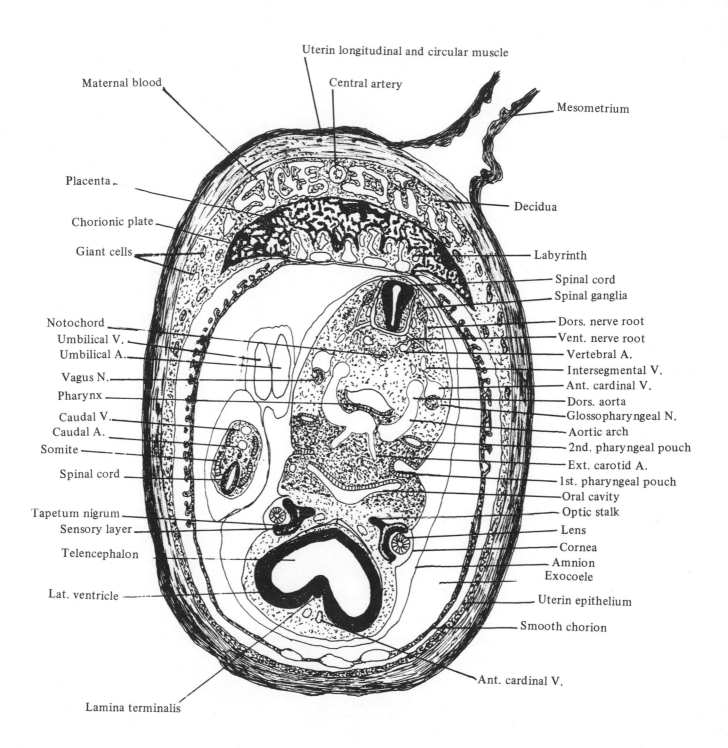

Uterin longitudinal and circular muscle

Central artery

Mesometrium

Maternal blood

Placenta

Decidua

Chorionic plate

Giant cells

Labyrinth

Spinal cord

Spinal ganglia

Notochord

Dors. nerve root

Umbilical V.

Vent. nerve root

Umbilical A.

Vertebral A.

Intersegmental V.

Vagus N.

Ant. cardinal V.

Pharynx

Dors. aorta

Caudal V.

Glossopharyngeal N.

Caudal A.

Aortic arch

Somite

2nd. pharyngeal pouch

Spinal cord

Ext. carotid A.

1st. pharyngeal pouch

Oral cavity

Tapetum nigrum

Optic stalk

Sensory layer

Lens

Telencephalon

Cornea

Amnion

Exocoele

Lat. ventricle

Uterin epithelium

Smooth chorion

Ant. cardinal V.

Lamina terminalis

Mammalian Development II
10 mm. Pig Embryo

10 mm. pig embryo
x 24, w.m.

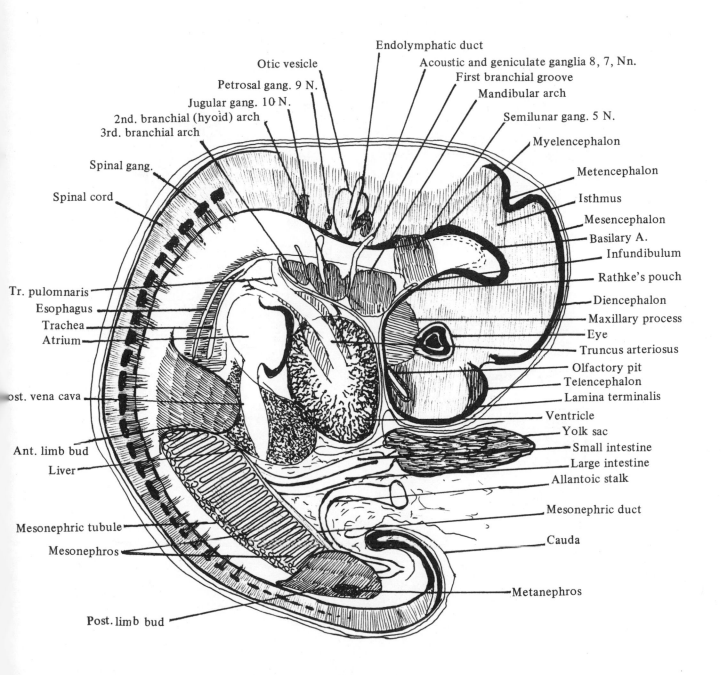

Endolymphatic duct
Otic vesicle
Acoustic and geniculate ganglia 8, 7, Nn.
Petrosal gang. 9 N.
First branchial groove
Jugular gang. 10 N.
Mandibular arch
2nd. branchial (hyoid) arch
Semilunar gang. 5 N.
3rd. branchial arch
Myelencephalon
Spinal gang.
Metencephalon
Spinal cord
Isthmus
Mesencephalon
Basilary A.
Infundibulum
Tr. pulomnaris
Rathke's pouch
Esophagus
Diencephalon
Trachea
Maxillary process
Atrium
Eye
Truncus arteriosus
Olfactory pit
Telencephalon
ost. vena cava
Lamina terminalis
Ventricle
Yolk sac
Ant. limb bud
Small intestine
Liver
Large intestine
Allantoic stalk
Mesonephric duct
Mesonephric tubule
Cauda
Mesonephros
Metanephros
Post. limb bud

237

10 mm. pig embryo
x 24, sections

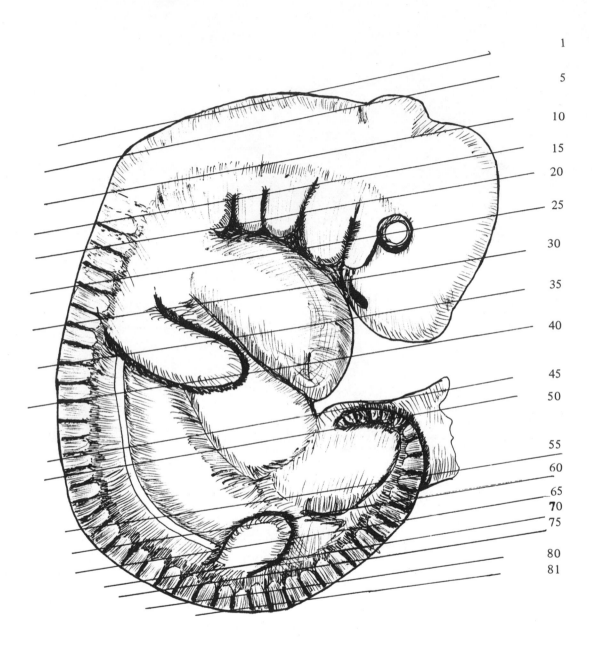

1

5

10

15

20

25

30

35

40

45

50

55

60

65

70

75

80

81

10 mm. pig embryo
x 40, x. sec., section 1

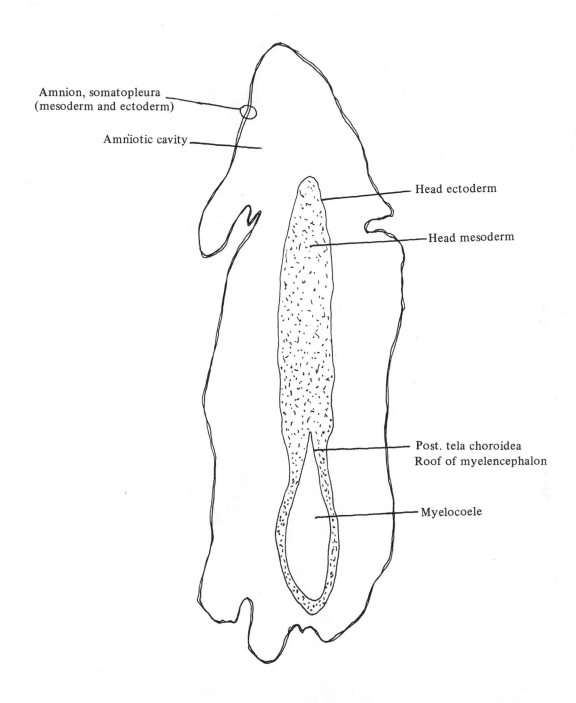

Amnion, somatopleura
(mesoderm and ectoderm)

Amniotic cavity

Head ectoderm

Head mesoderm

Post. tela choroidea
Roof of myelencephalon

Myelocoele

10 mm. pig embryo
x 40, x. sec., section 2

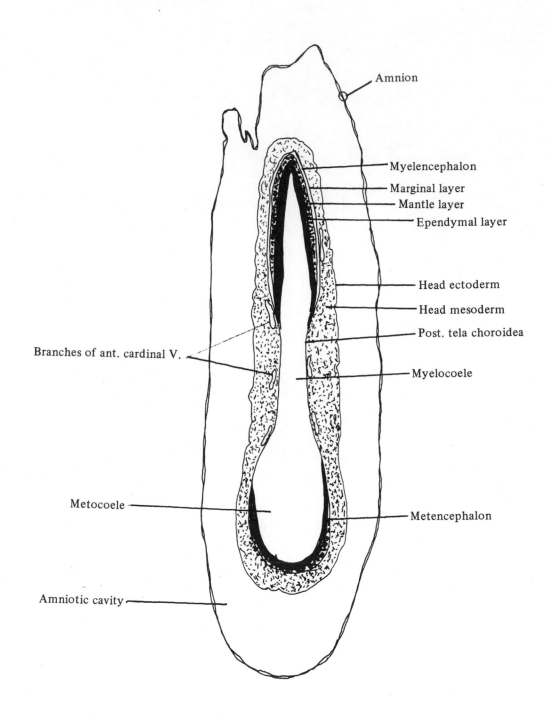

Amnion

Myelencephalon

Marginal layer

Mantle layer

Ependymal layer

Head ectoderm

Head mesoderm

Post. tela choroidea

Branches of ant. cardinal V.

Myelocoele

Metocoele

Metencephalon

Amniotic cavity

240

10 mm. pig embryo
x 40, x. sec., section 3

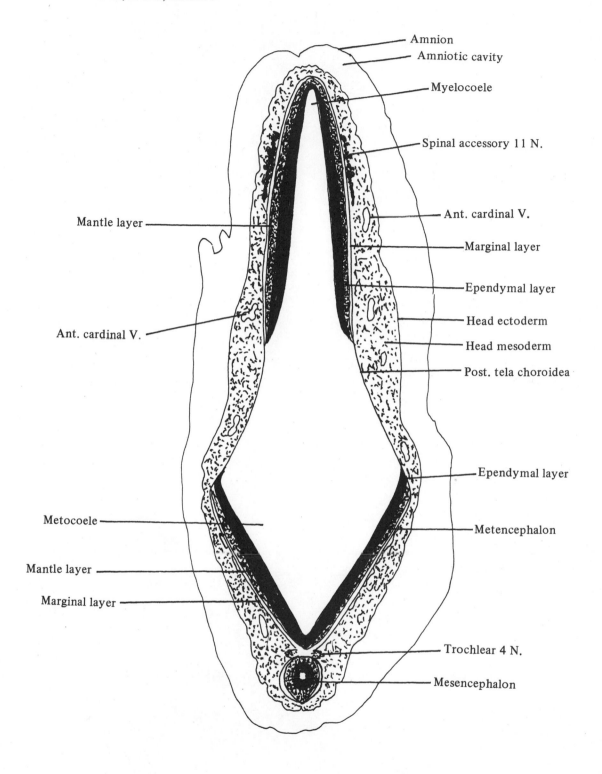

Amnion

Amniotic cavity

Myelocoele

Spinal accessory 11 N.

Mantle layer

Ant. cardinal V.

Marginal layer

Ependymal layer

Head ectoderm

Head mesoderm

Ant. cardinal V.

Post. tela choroidea

Ependymal layer

Metocoele

Metencephalon

Mantle layer

Marginal layer

Trochlear 4 N.

Mesencephalon

10 mm. pig embryo
x 40, x. sec, section 4

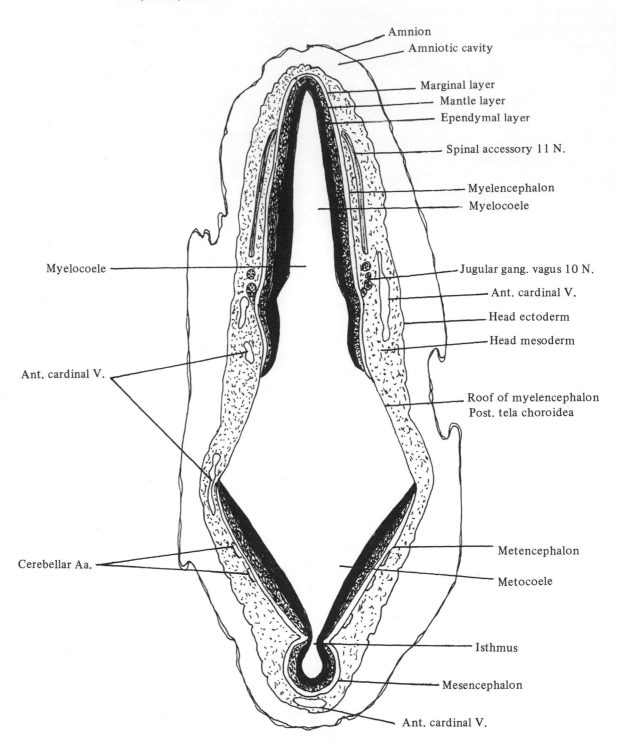

Amnion
Amniotic cavity

Marginal layer
Mantle layer
Ependymal layer

Spinal accessory 11 N.

Myelencephalon
Myelocoele

Myelocoele

Jugular gang. vagus 10 N.

Ant. cardinal V.

Head ectoderm

Head mesoderm

Ant. cardinal V.

Roof of myelencephalon
Post. tela choroidea

Cerebellar Aa.

Metencephalon

Metocoele

Isthmus

Mesencephalon

Ant. cardinal V.

10 mm. pig embryo
x 40, x. sec. section 5

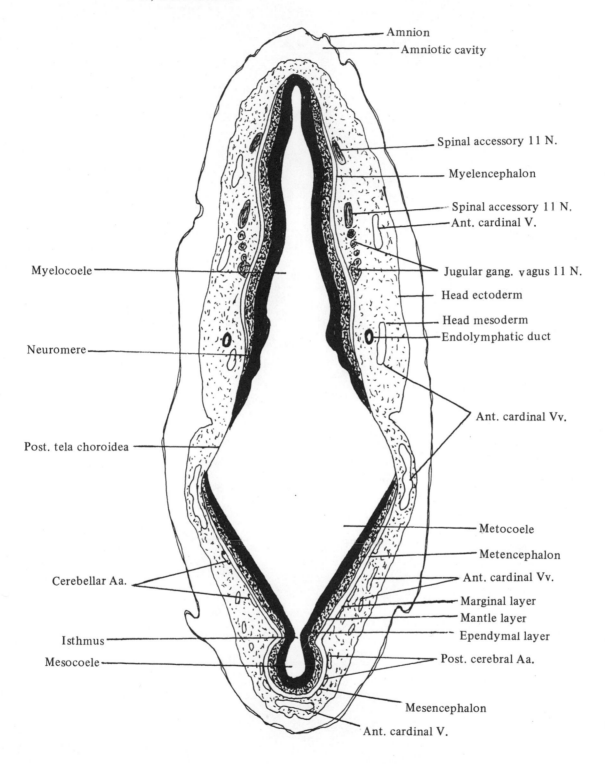

Amnion

Amniotic cavity

Spinal accessory 11 N.

Myelencephalon

Spinal accessory 11 N.
Ant. cardinal V.

Jugular gang. vagus 11 N.

Head ectoderm

Head mesoderm
Endolymphatic duct

Ant. cardinal Vv.

Metocoele

Metencephalon

Ant. cardinal Vv.

Marginal layer

Mantle layer

Ependymal layer

Post. cerebral Aa.

Mesencephalon

Ant. cardinal V.

Myelocoele

Neuromere

Post. tela choroidea

Cerebellar Aa.

Isthmus

Mesocoele

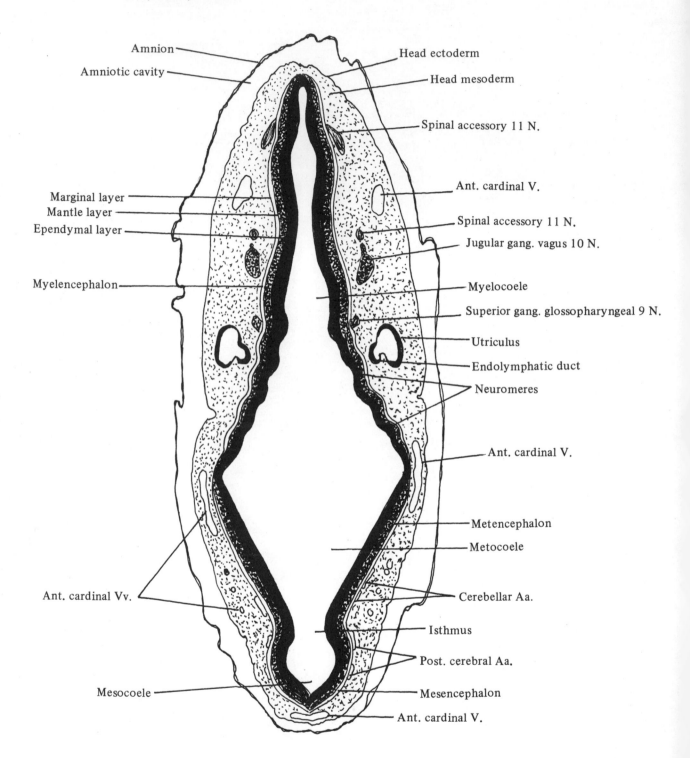

Amnion

Amniotic cavity

Head ectoderm

Head mesoderm

Spinal accessory 11 N.

Ant. cardinal V.

Marginal layer

Mantle layer

Ependymal layer

Spinal accessory 11 N.

Jugular gang. vagus 10 N.

Myelencephalon

Myelocoele

Superior gang. glossopharyngeal 9 N.

Utriculus

Endolymphatic duct

Neuromeres

Ant. cardinal V.

Metencephalon

Metocoele

Ant. cardinal Vv.

Cerebellar Aa.

Isthmus

Post. cerebral Aa.

Mesocoele

Mesencephalon

Ant. cardinal V.

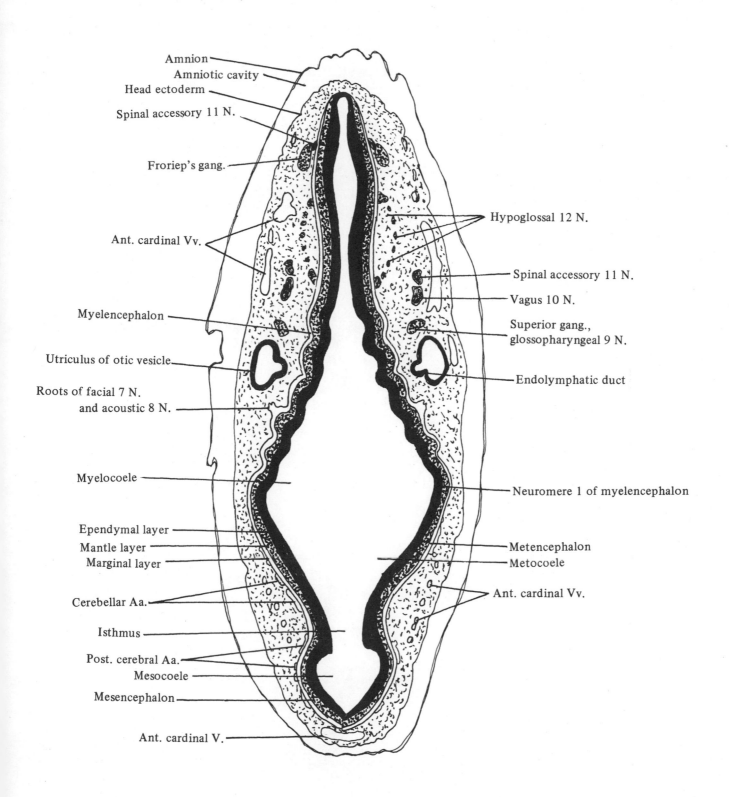

Amnion

Amniotic cavity

Head ectoderm

Spinal accessory 11 N.

Froriep's gang.

Ant. cardinal Vv.

Myelencephalon

Utriculus of otic vesicle

Roots of facial 7 N.
and acoustic 8 N.

Myelocoele

Ependymal layer

Mantle layer

Marginal layer

Cerebellar Aa.

Isthmus

Post. cerebral Aa.

Mesocoele

Mesencephalon

Ant. cardinal V.

Hypoglossal 12 N.

Spinal accessory 11 N.

Vagus 10 N.

Superior gang.,
glossopharyngeal 9 N.

Endolymphatic duct

Neuromere 1 of myelencephalon

Metencephalon

Metocoele

Ant. cardinal Vv.

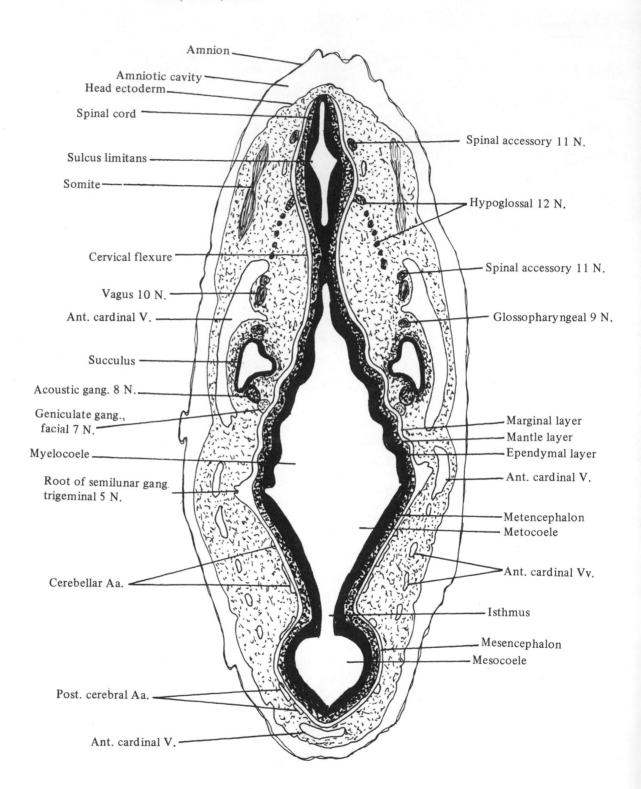

Amnion

Amniotic cavity
Head ectoderm

Spinal cord

Spinal accessory 11 N.

Sulcus limitans

Somite

Hypoglossal 12 N.

Cervical flexure

Spinal accessory 11 N.

Vagus 10 N.
Ant. cardinal V.

Glossopharyngeal 9 N.

Succulus

Acoustic gang. 8 N.

Geniculate gang.,
facial 7 N.

Marginal layer
Mantle layer
Ependymal layer

Myelocoele

Ant. cardinal V.

Root of semilunar gang.
trigeminal 5 N.

Metencephalon
Metocoele

Ant. cardinal Vv.

Cerebellar Aa.

Isthmus

Mesencephalon
Mesocoele

Post. cerebral Aa.

Ant. cardinal V.

10 mm. pig embryo
x 40, x. sec., section 9

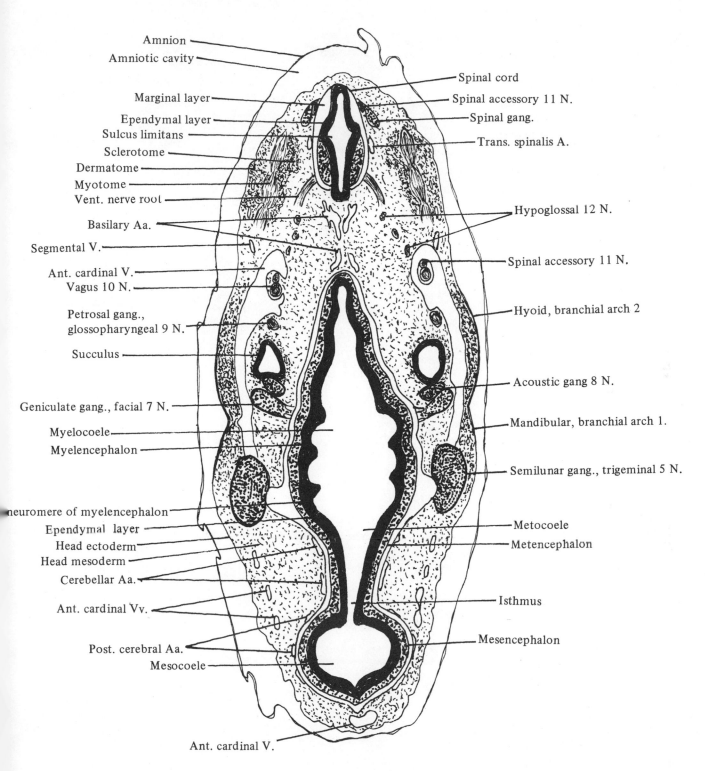

Amnion

Amniotic cavity

Marginal layer

Ependymal layer

Sulcus limitans

Sclerotome

Dermatome

Myotome

Vent. nerve root

Basilary Aa.

Segmental V.

Ant. cardinal V.

Vagus 10 N.

Petrosal gang.,
glossopharyngeal 9 N.

Succulus

Geniculate gang., facial 7 N.

Myelocoele

Myelencephalon

neuromere of myelencephalon

Ependymal layer

Head ectoderm

Head mesoderm

Cerebellar Aa.

Ant. cardinal Vv.

Post. cerebral Aa.

Mesocoele

Ant. cardinal V.

Spinal cord

Spinal accessory 11 N.

Spinal gang.

Trans. spinalis A.

Hypoglossal 12 N.

Spinal accessory 11 N.

Hyoid, branchial arch 2

Acoustic gang 8 N.

Mandibular, branchial arch 1.

Semilunar gang., trigeminal 5 N.

Metocoele

Metencephalon

Isthmus

Mesencephalon

10 mm. pig embryo
x 40, x. sec. section 10

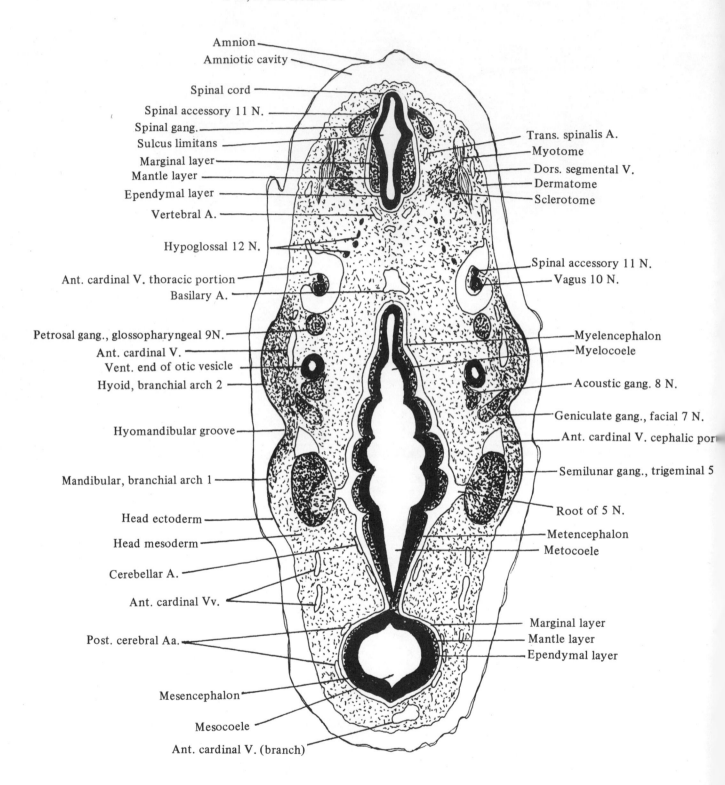

Amnion
Amniotic cavity
Spinal cord
Spinal accessory 11 N.
Spinal gang.
Sulcus limitans
Marginal layer
Mantle layer
Ependymal layer
Vertebral A.
Hypoglossal 12 N.
Ant. cardinal V. thoracic portion
Basilary A.
Petrosal gang., glossopharyngeal 9N.
Ant. cardinal V.
Vent. end of otic vesicle
Hyoid, branchial arch 2
Hyomandibular groove
Mandibular, branchial arch 1
Head ectoderm
Head mesoderm
Cerebellar A.
Ant. cardinal Vv.
Post. cerebral Aa.
Mesencephalon
Mesocoele
Ant. cardinal V. (branch)

Trans. spinalis A.
Myotome
Dors. segmental V.
Dermatome
Sclerotome

Spinal accessory 11 N.
Vagus 10 N.

Myelencephalon
Myelocoele

Acoustic gang. 8 N.

Geniculate gang., facial 7 N.
Ant. cardinal V. cephalic por

Semilunar gang., trigeminal 5

Root of 5 N.
Metencephalon
Metocoele

Marginal layer
Mantle layer
Ependymal layer

10 mm. pig embryo
x 40, x. sec, section 11

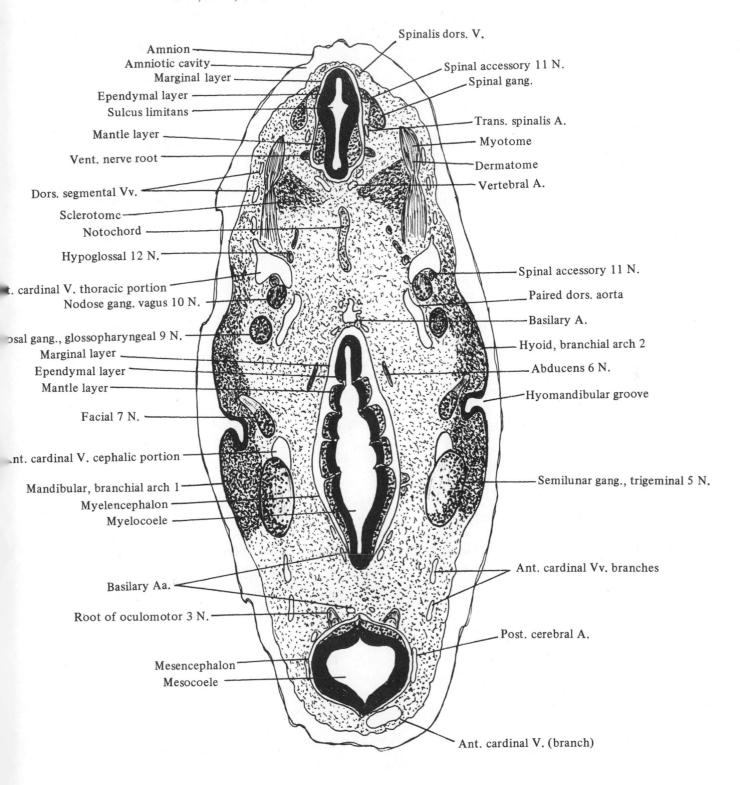

Spinalis dors. V.

Amnion
Amniotic cavity
Marginal layer
Ependymal layer
Sulcus limitans

Spinal accessory 11 N.
Spinal gang.

Mantle layer
Vent. nerve root

Trans. spinalis A.
Myotome
Dermatome

Dors. segmental Vv.

Vertebral A.

Sclerotome
Notochord
Hypoglossal 12 N.
t. cardinal V. thoracic portion
Nodose gang. vagus 10 N.
osal gang., glossopharyngeal 9 N.
Marginal layer
Ependymal layer
Mantle layer

Spinal accessory 11 N.
Paired dors. aorta
Basilary A.
Hyoid, branchial arch 2
Abducens 6 N.
Hyomandibular groove

Facial 7 N.

nt. cardinal V. cephalic portion
Mandibular, branchial arch 1
Myelencephalon
Myelocoele

Semilunar gang., trigeminal 5 N.

Ant. cardinal Vv. branches

Basilary Aa.
Root of oculomotor 3 N.

Post. cerebral A.

Mesencephalon
Mesocoele

Ant. cardinal V. (branch)

249

10 mm. pig embryo
x 40, x. sec., section 12

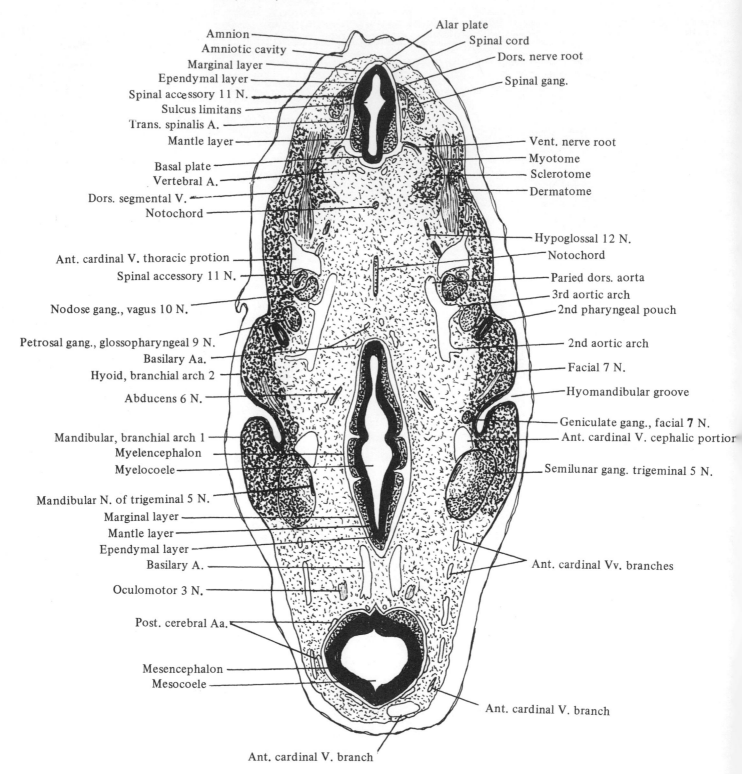

Amnion
Amniotic cavity
Marginal layer
Ependymal layer
Spinal accessory 11 N.
Sulcus limitans
Trans. spinalis A.
Mantle layer
Basal plate
Vertebral A.
Dors. segmental V.
Notochord

Ant. cardinal V. thoracic protion
Spinal accessory 11 N.

Nodose gang., vagus 10 N.

Petrosal gang., glossopharyngeal 9 N.
Basilary Aa.
Hyoid, branchial arch 2
Abducens 6 N.

Mandibular, branchial arch 1
Myelencephalon
Myelocoele

Mandibular N. of trigeminal 5 N.
Marginal layer
Mantle layer
Ependymal layer
Basilary A.

Oculomotor 3 N.

Post. cerebral Aa.

Mesencephalon
Mesocoele

Alar plate
Spinal cord
Dors. nerve root
Spinal gang.

Vent. nerve root
Myotome
Sclerotome
Dermatome

Hypoglossal 12 N.
Notochord
Paried dors. aorta
3rd aortic arch
2nd pharyngeal pouch

2nd aortic arch
Facial 7 N.
Hyomandibular groove
Geniculate gang., facial 7 N.
Ant. cardinal V. cephalic portion
Semilunar gang. trigeminal 5 N.

Ant. cardinal Vv. branches

Ant. cardinal V. branch

Ant. cardinal V. branch

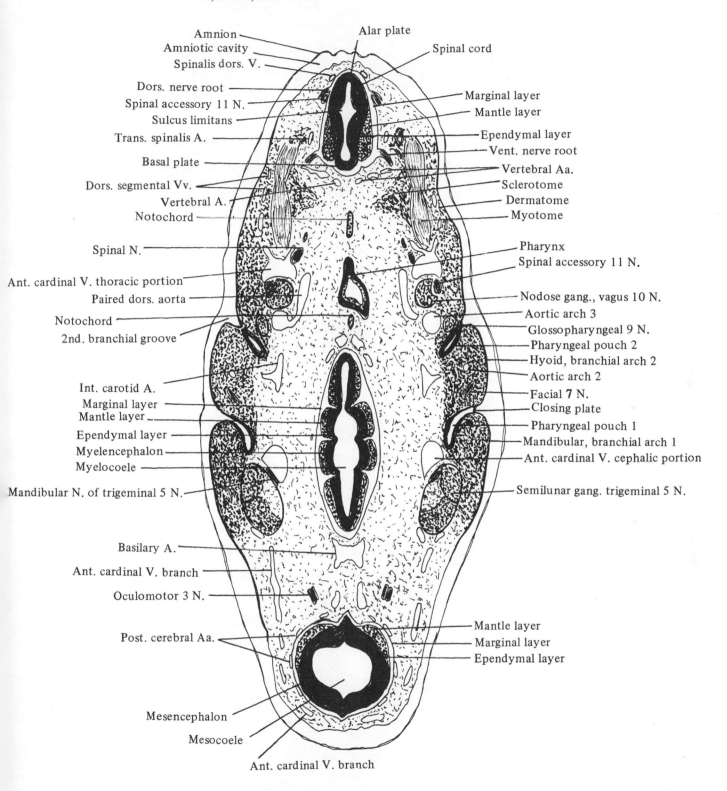

Amnion
Amniotic cavity
Spinalis dors. V.
Alar plate
Spinal cord

Dors. nerve root
Spinal accessory 11 N.
Sulcus limitans
Trans. spinalis A.
Basal plate
Dors. segmental Vv.
Vertebral A.
Notochord
Marginal layer
Mantle layer
Ependymal layer
Vent. nerve root
Vertebral Aa.
Sclerotome
Dermatome
Myotome

Spinal N.
Ant. cardinal V. thoracic portion
Paired dors. aorta
Notochord
2nd. branchial groove
Pharynx
Spinal accessory 11 N.
Nodose gang., vagus 10 N.
Aortic arch 3
Glossopharyngeal 9 N.
Pharyngeal pouch 2
Hyoid, branchial arch 2
Aortic arch 2

Int. carotid A.
Marginal layer
Mantle layer
Ependymal layer
Myelencephalon
Myelocoele
Mandibular N. of trigeminal 5 N.
Facial 7 N.
Closing plate
Pharyngeal pouch 1
Mandibular, branchial arch 1
Ant. cardinal V. cephalic portion
Semilunar gang. trigeminal 5 N.

Basilary A.
Ant. cardinal V. branch
Oculomotor 3 N.
Mantle layer
Marginal layer
Ependymal layer

Post. cerebral Aa.

Mesencephalon
Mesocoele
Ant. cardinal V. branch

10 mm. pig embryo
x 49, x. sec., section 14

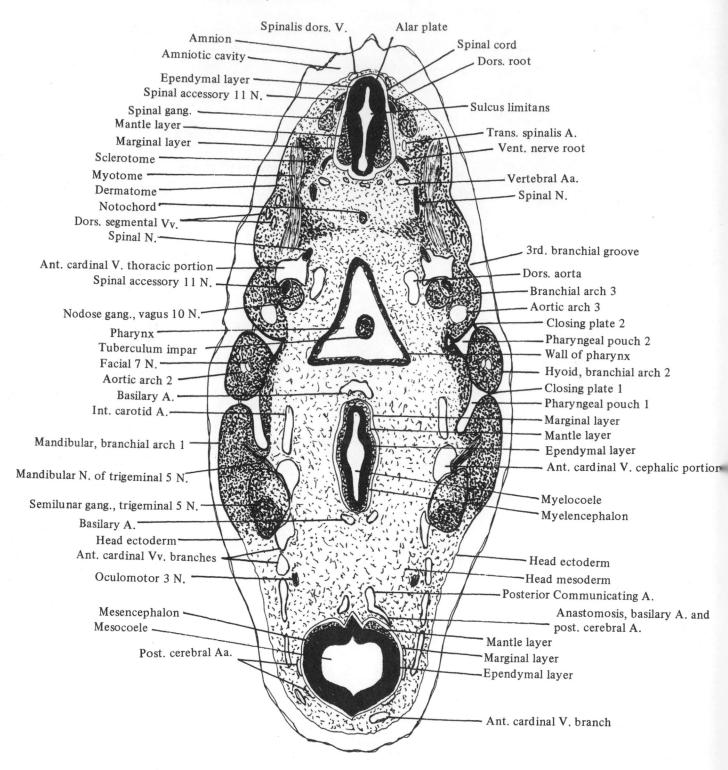

Spinalis dors. V.
Alar plate
Amnion
Spinal cord
Amniotic cavity
Dors. root
Ependymal layer
Spinal accessory 11 N.
Sulcus limitans
Spinal gang.
Mantle layer
Marginal layer
Trans. spinalis A.
Vent. nerve root
Sclerotome
Myotome
Vertebral Aa.
Dermatome
Spinal N.
Notochord
Dors. segmental Vv.
Spinal N.
3rd. branchial groove
Ant. cardinal V. thoracic portion
Dors. aorta
Spinal accessory 11 N.
Branchial arch 3
Aortic arch 3
Nodose gang., vagus 10 N.
Closing plate 2
Pharynx
Pharyngeal pouch 2
Tuberculum impar
Wall of pharynx
Facial 7 N.
Hyoid, branchial arch 2
Aortic arch 2
Closing plate 1
Basilary A.
Pharyngeal pouch 1
Int. carotid A.
Marginal layer
Mantle layer
Mandibular, branchial arch 1
Ependymal layer
Ant. cardinal V. cephalic portion
Mandibular N. of trigeminal 5 N.
Semilunar gang., trigeminal 5 N.
Myelocoele
Basilary A.
Myelencephalon
Head ectoderm
Ant. cardinal Vv. branches
Oculomotor 3 N.
Head ectoderm
Head mesoderm
Posterior Communicating A.
Mesencephalon
Mesocoele
Anastomosis, basilary A. and
post. cerebral A.
Mantle layer
Post. cerebral Aa.
Marginal layer
Ependymal layer

Ant. cardinal V. branch

252

10 mm. pig embryo
x 40, x. sec., section 15

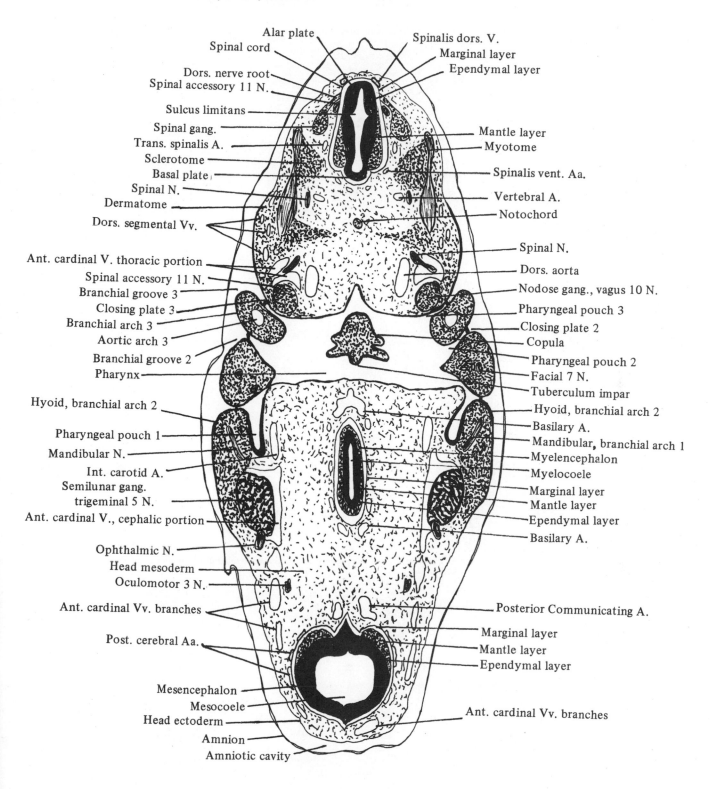

Alar plate
Spinal cord
Dors. nerve root
Spinal accessory 11 N.
Sulcus limitans
Spinal gang.
Trans. spinalis A.
Sclerotome
Basal plate
Spinal N.
Dermatome
Dors. segmental Vv.
Ant. cardinal V. thoracic portion
Spinal accessory 11 N.
Branchial groove 3
Closing plate 3
Branchial arch 3
Aortic arch 3
Branchial groove 2
Pharynx
Hyoid, branchial arch 2
Pharyngeal pouch 1
Mandibular N.
Int. carotid A.
Semilunar gang.
trigeminal 5 N.
Ant. cardinal V., cephalic portion
Ophthalmic N.
Head mesoderm
Oculomotor 3 N.
Ant. cardinal Vv. branches
Post. cerebral Aa.
Mesencephalon
Mesocoele
Head ectoderm
Amnion
Amniotic cavity

Spinalis dors. V.
Marginal layer
Ependymal layer
Mantle layer
Myotome
Spinalis vent. Aa.
Vertebral A.
Notochord
Spinal N.
Dors. aorta
Nodose gang., vagus 10 N.
Pharyngeal pouch 3
Closing plate 2
Copula
Pharyngeal pouch 2
Facial 7 N.
Tuberculum impar
Hyoid, branchial arch 2
Basilary A.
Mandibular, branchial arch 1
Myelencephalon
Myelocoele
Marginal layer
Mantle layer
Ependymal layer
Basilary A.
Posterior Communicating A.
Marginal layer
Mantle layer
Ependymal layer
Ant. cardinal Vv. branches

253

10 mm. pig embryo
x 40, x. sec., section 16

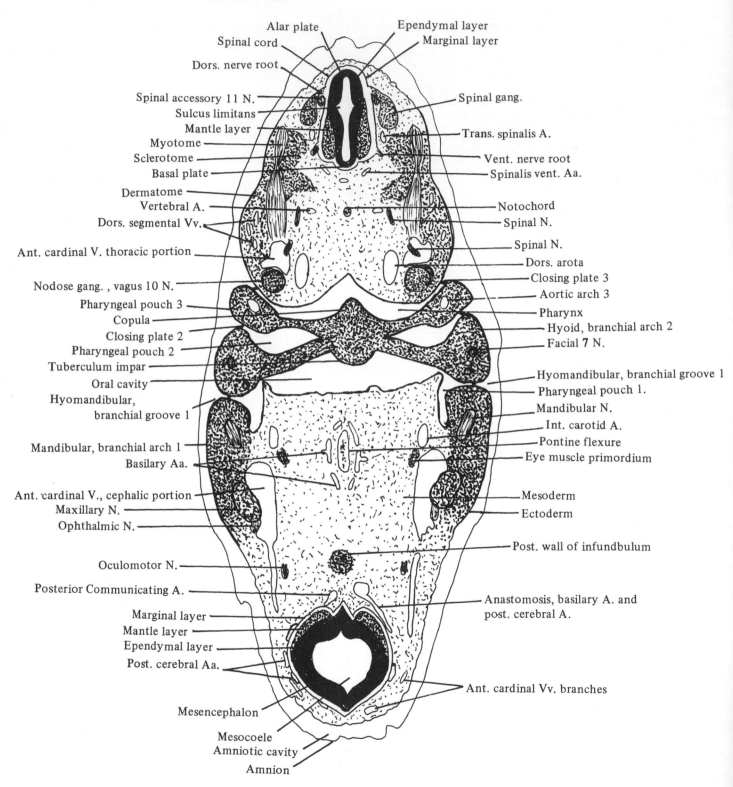

Alar plate
Spinal cord
Dors. nerve root
Spinal accessory 11 N.
Sulcus limitans
Mantle layer
Myotome
Sclerotome
Basal plate
Dermatome
Vertebral A.
Dors. segmental Vv.
Ant. cardinal V. thoracic portion
Nodose gang. , vagus 10 N.
Pharyngeal pouch 3
Copula
Closing plate 2
Pharyngeal pouch 2
Tuberculum impar
Oral cavity
Hyomandibular,
branchial groove 1
Mandibular, branchial arch 1
Basilary Aa.
Ant. cardinal V., cephalic portion
Maxillary N.
Ophthalmic N.
Oculomotor N.
Posterior Communicating A.
Marginal layer
Mantle layer
Ependymal layer
Post. cerebral Aa.
Mesencephalon
Mesocoele
Amniotic cavity
Amnion

Ependymal layer
Marginal layer
Spinal gang.
Trans. spinalis A.
Vent. nerve root
Spinalis vent. Aa.
Notochord
Spinal N.
Spinal N.
Dors. arota
Closing plate 3
Aortic arch 3
Pharynx
Hyoid, branchial arch 2
Facial 7 N.
Hyomandibular, branchial groove 1
Pharyngeal pouch 1.
Mandibular N.
Int. carotid A.
Pontine flexure
Eye muscle primordium
Mesoderm
Ectoderm
Post. wall of infundbulum
Anastomosis, basilary A. and
post. cerebral A.
Ant. cardinal Vv. branches

10 mm. pig embryo
x 40, x. sec., section 17

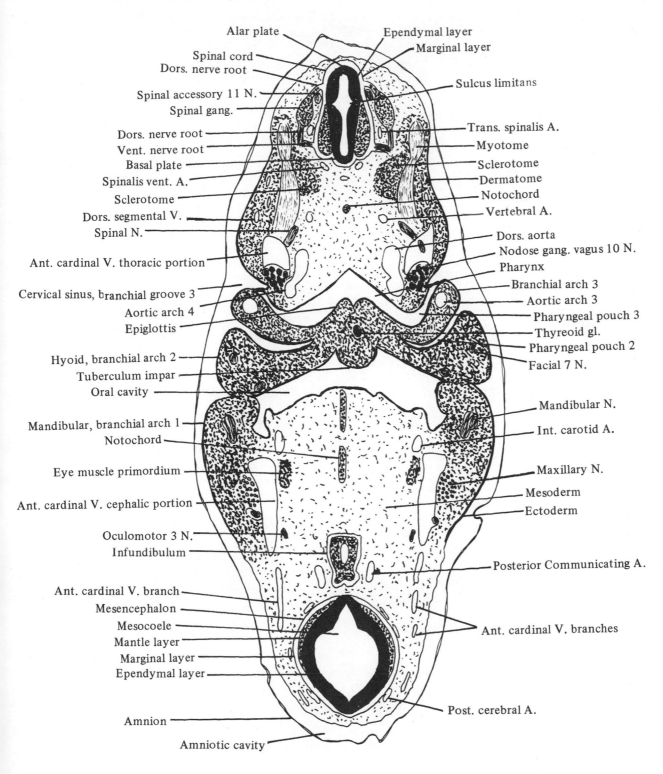

Alar plate

Ependymal layer

Marginal layer

Spinal cord

Dors. nerve root

Sulcus limitans

Spinal accessory 11 N.

Spinal gang.

Dors. nerve root

Trans. spinalis A.

Vent. nerve root

Myotome

Basal plate

Sclerotome

Spinalis vent. A.

Dermatome

Sclerotome

Notochord

Dors. segmental V.

Vertebral A.

Spinal N.

Dors. aorta

Ant. cardinal V. thoracic portion

Nodose gang. vagus 10 N.

Pharynx

Cervical sinus, branchial groove 3

Branchial arch 3

Aortic arch 4

Aortic arch 3

Epiglottis

Pharyngeal pouch 3

Thyreoid gl.

Hyoid, branchial arch 2

Pharyngeal pouch 2

Tuberculum impar

Facial 7 N.

Oral cavity

Mandibular N.

Mandibular, branchial arch 1

Int. carotid A.

Notochord

Eye muscle primordium

Maxillary N.

Ant. cardinal V. cephalic portion

Mesoderm

Ectoderm

Oculomotor 3 N.

Infundibulum

Posterior Communicating A.

Ant. cardinal V. branch

Mesencephalon

Mesocoele

Mantle layer

Ant. cardinal V. branches

Marginal layer

Ependymal layer

Amnion

Post. cerebral A.

Amniotic cavity

10 mm. pig embryo
x 40, x. sec., section 18

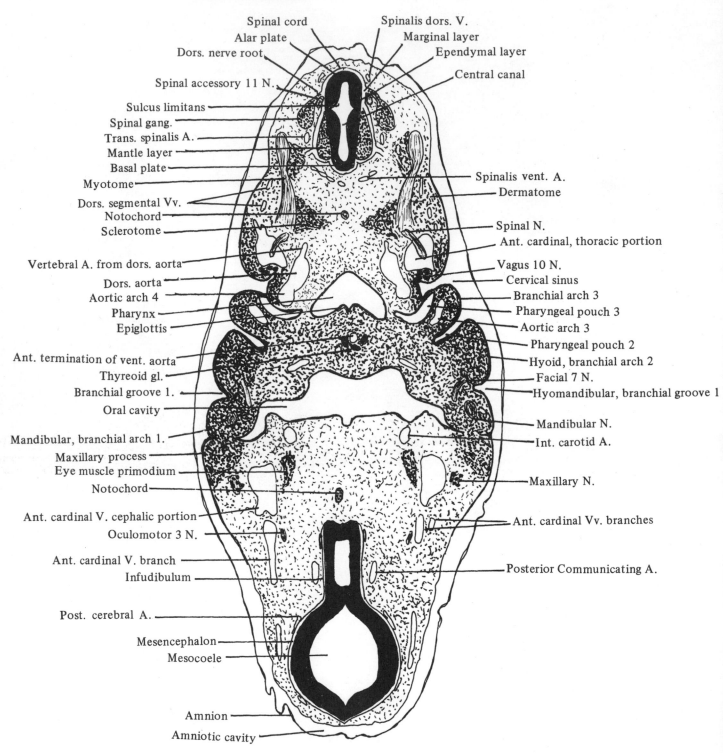

Spinal cord
Alar plate
Dors. nerve root
Spinal accessory 11 N.
Sulcus limitans
Spinal gang.
Trans. spinalis A.
Mantle layer
Basal plate
Myotome
Dors. segmental Vv.
Notochord
Sclerotome
Vertebral A. from dors. aorta
Dors. aorta
Aortic arch 4
Pharynx
Epiglottis
Ant. termination of vent. aorta
Thyreoid gl.
Branchial groove 1.
Oral cavity
Mandibular, branchial arch 1.
Maxillary process
Eye muscle primodium
Notochord
Ant. cardinal V. cephalic portion
Oculomotor 3 N.
Ant. cardinal V. branch
Infudibulum
Post. cerebral A.
Mesencephalon
Mesocoele
Amnion
Amniotic cavity

Spinalis dors. V.
Marginal layer
Ependymal layer
Central canal
Spinalis vent. A.
Dermatome
Spinal N.
Ant. cardinal, thoracic portion
Vagus 10 N.
Cervical sinus
Branchial arch 3
Pharyngeal pouch 3
Aortic arch 3
Pharyngeal pouch 2
Hyoid, branchial arch 2
Facial 7 N.
Hyomandibular, branchial groove 1
Mandibular N.
Int. carotid A.
Maxillary N.
Ant. cardinal Vv. branches
Posterior Communicating A.

10 mm. pig embryo
x 40, x. sec., section 19

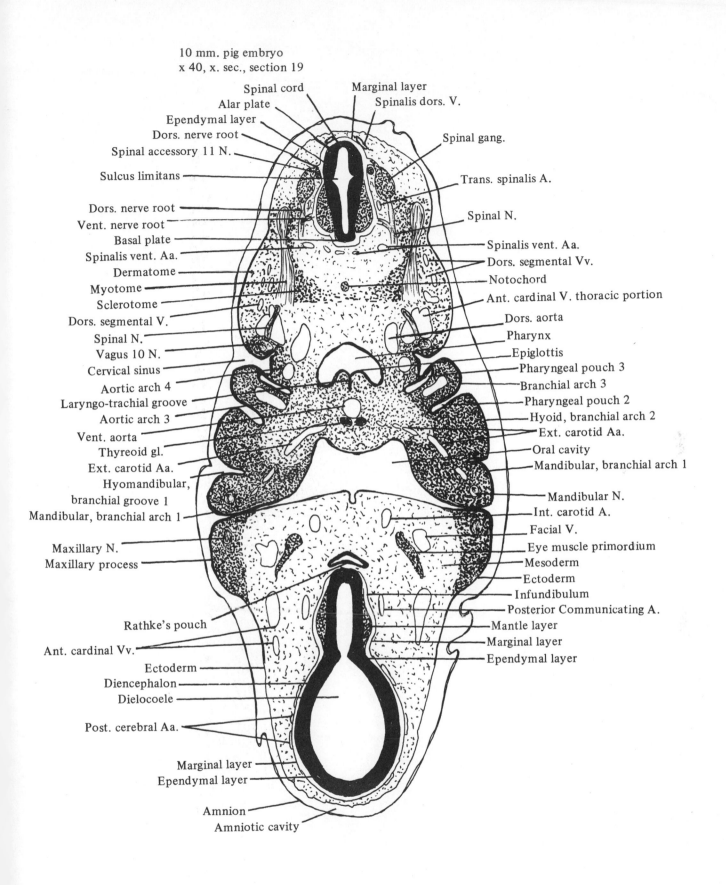

Spinal cord
Alar plate
Ependymal layer
Dors. nerve root
Spinal accessory 11 N.
Sulcus limitans
Dors. nerve root
Vent. nerve root
Basal plate
Spinalis vent. Aa.
Dermatome
Myotome
Sclerotome
Dors. segmental V.
Spinal N.
Vagus 10 N.
Cervical sinus
Aortic arch 4
Laryngo-trachial groove
Aortic arch 3
Vent. aorta
Thyreoid gl.
Ext. carotid Aa.
Hyomandibular,
branchial groove 1
Mandibular, branchial arch 1
Maxillary N.
Maxillary process
Rathke's pouch
Ant. cardinal Vv.
Ectoderm
Diencephalon
Dielocoele
Post. cerebral Aa.
Marginal layer
Ependymal layer
Amnion
Amniotic cavity

Marginal layer
Spinalis dors. V.
Spinal gang.
Trans. spinalis A.
Spinal N.
Spinalis vent. Aa.
Dors. segmental Vv.
Notochord
Ant. cardinal V. thoracic portion
Dors. aorta
Pharynx
Epiglottis
Pharyngeal pouch 3
Branchial arch 3
Pharyngeal pouch 2
Hyoid, branchial arch 2
Ext. carotid Aa.
Oral cavity
Mandibular, branchial arch 1
Mandibular N.
Int. carotid A.
Facial V.
Eye muscle primordium
Mesoderm
Ectoderm
Infundibulum
Posterior Communicating A.
Mantle layer
Marginal layer
Ependymal layer

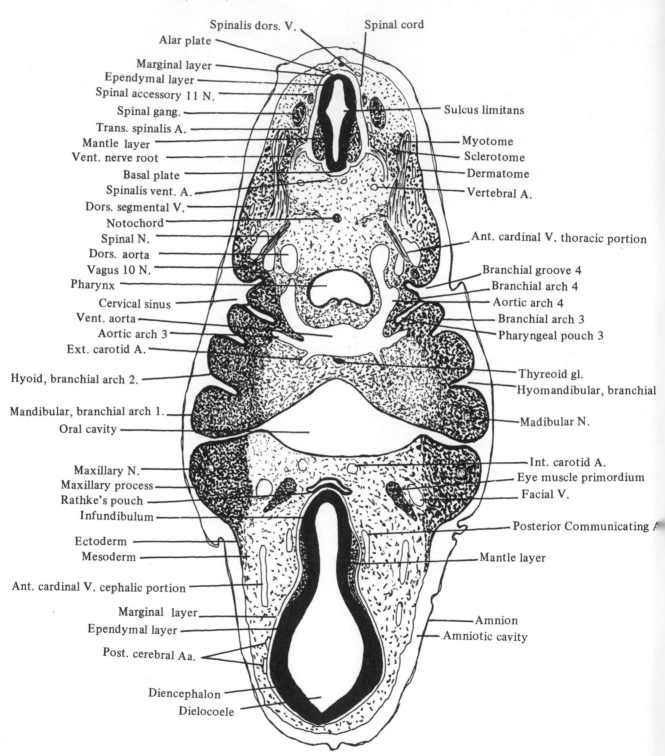

10 mm. pig embryo
x 40, x. sec., section 20

Spinalis dors. V.
Spinal cord
Alar plate

Marginal layer
Ependymal layer
Spinal accessory 11 N.
Spinal gang.
Trans. spinalis A.
Mantle layer
Vent. nerve root
Basal plate
Spinalis vent. A.
Dors. segmental V.
Notochord
Spinal N.
Dors. aorta
Vagus 10 N.
Pharynx
Cervical sinus
Vent. aorta
Aortic arch 3
Ext. carotid A.
Hyoid, branchial arch 2.
Mandibular, branchial arch 1.
Oral cavity
Maxillary N.
Maxillary process
Rathke's pouch
Infundibulum
Ectoderm
Mesoderm
Ant. cardinal V. cephalic portion
Marginal layer
Ependymal layer
Post. cerebral Aa.
Diencephalon
Dielocoele

Sulcus limitans
Myotome
Sclerotome
Dermatome
Vertebral A.
Ant. cardinal V. thoracic portion
Branchial groove 4
Branchial arch 4
Aortic arch 4
Branchial arch 3
Pharyngeal pouch 3
Thyreoid gl.
Hyomandibular, branchial
Madibular N.
Int. carotid A.
Eye muscle primordium
Facial V.
Posterior Communicating A.
Mantle layer
Amnion
Amniotic cavity

10 mm. pig embryo
x 40, x. sec. section 21

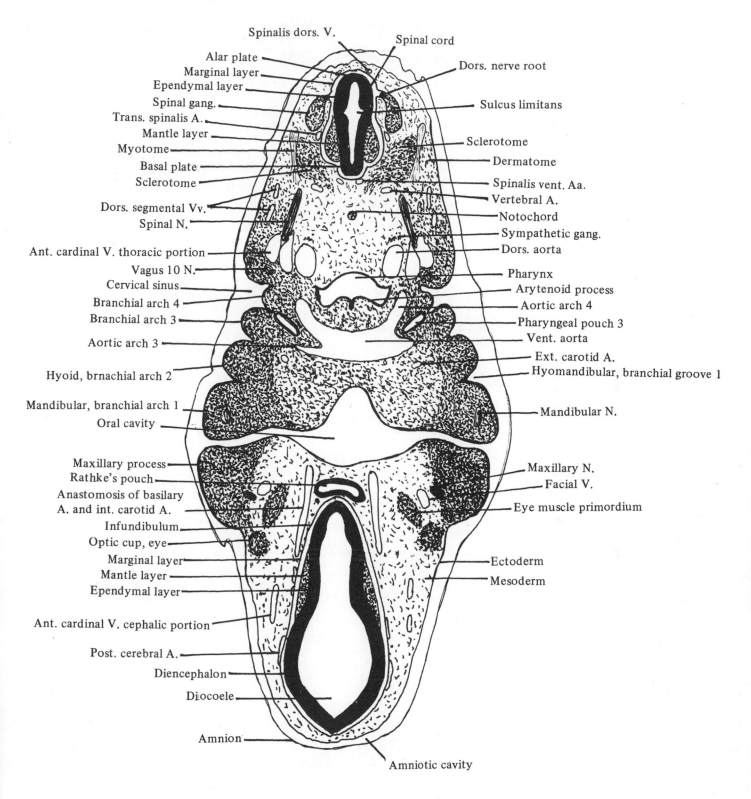

Spinalis dors. V.
Spinal cord
Alar plate
Marginal layer
Ependymal layer
Spinal gang.
Trans. spinalis A.
Mantle layer
Myotome
Basal plate
Sclerotome
Dors. segmental Vv.
Spinal N.
Ant. cardinal V. thoracic portion
Vagus 10 N.
Cervical sinus
Branchial arch 4
Branchial arch 3
Aortic arch 3
Hyoid, brnachial arch 2
Mandibular, branchial arch 1
Oral cavity
Maxillary process
Rathke's pouch
Anastomosis of basilary
A. and int. carotid A.
Infundibulum
Optic cup, eye
Marginal layer
Mantle layer
Ependymal layer
Ant. cardinal V. cephalic portion
Post. cerebral A.
Diencephalon
Diocoele
Amnion

Dors. nerve root
Sulcus limitans
Sclerotome
Dermatome
Spinalis vent. Aa.
Vertebral A.
Notochord
Sympathetic gang.
Dors. aorta
Pharynx
Arytenoid process
Aortic arch 4
Pharyngeal pouch 3
Vent. aorta
Ext. carotid A.
Hyomandibular, branchial groove 1
Mandibular N.
Maxillary N.
Facial V.
Eye muscle primordium
Ectoderm
Mesoderm

Amniotic cavity

10 mm. pig embryo
x 40, x. sec., section 22

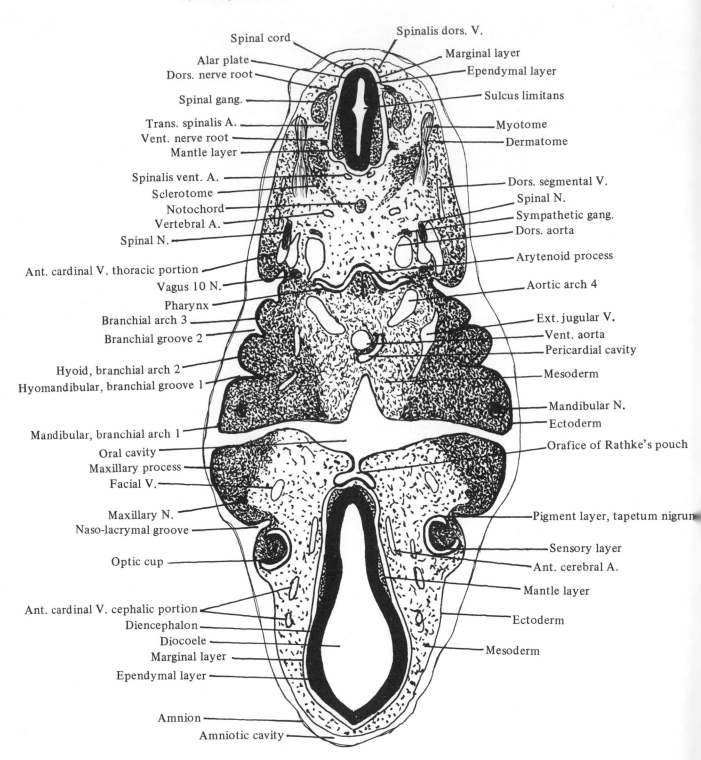

Spinal cord
Alar plate
Dors. nerve root
Spinal gang.
Trans. spinalis A.
Vent. nerve root
Mantle layer
Spinalis vent. A.
Sclerotome
Notochord
Vertebral A.
Spinal N.
Ant. cardinal V. thoracic portion
Vagus 10 N.
Pharynx
Branchial arch 3
Branchial groove 2
Hyoid, branchial arch 2
Hyomandibular, branchial groove 1
Mandibular, branchial arch 1
Oral cavity
Maxillary process
Facial V.
Maxillary N.
Naso-lacrymal groove
Optic cup
Ant. cardinal V. cephalic portion
Diencephalon
Diocoele
Marginal layer
Ependymal layer
Amnion
Amniotic cavity

Spinalis dors. V.
Marginal layer
Ependymal layer
Sulcus limitans
Myotome
Dermatome
Dors. segmental V.
Spinal N.
Sympathetic gang.
Dors. aorta
Arytenoid process
Aortic arch 4
Ext. jugular V.
Vent. aorta
Pericardial cavity
Mesoderm
Mandibular N.
Ectoderm
Orafice of Rathke's pouch
Pigment layer, tapetum nigrum
Sensory layer
Ant. cerebral A.
Mantle layer
Ectoderm
Mesoderm

10 mm. pig embryo
x 40, x. sec., section 23

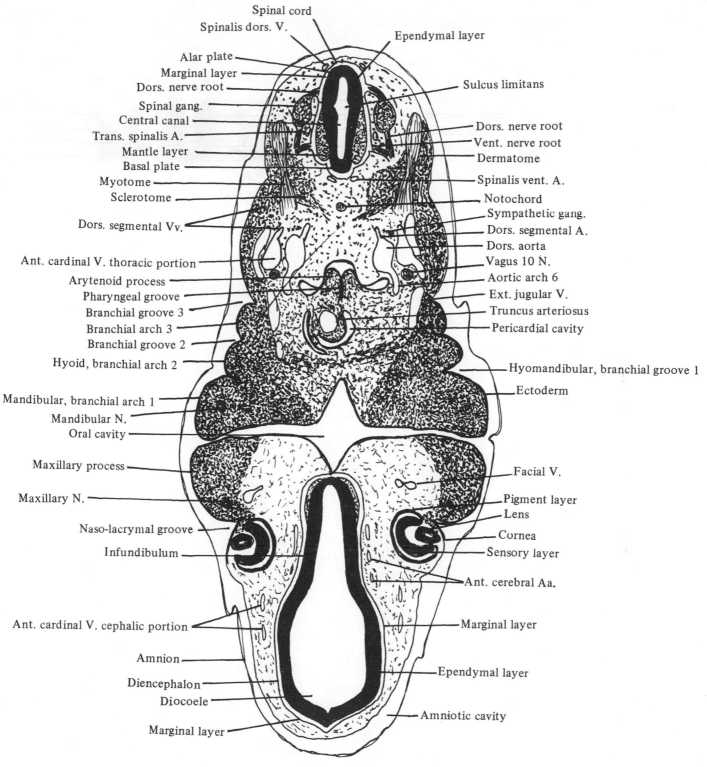

Spinal cord
Spinalis dors. V.
Alar plate
Marginal layer
Dors. nerve root
Spinal gang.
Central canal
Trans. spinalis A.
Mantle layer
Basal plate
Myotome
Sclerotome
Dors. segmental Vv.
Ant. cardinal V. thoracic portion
Arytenoid process
Pharyngeal groove
Branchial groove 3
Branchial arch 3
Branchial groove 2
Hyoid, branchial arch 2
Mandibular, branchial arch 1
Mandibular N.
Oral cavity
Maxillary process
Maxillary N.
Naso-lacrymal groove
Infundibulum
Ant. cardinal V. cephalic portion
Amnion
Diencephalon
Diocoele
Marginal layer

Ependymal layer
Sulcus limitans
Dors. nerve root
Vent. nerve root
Dermatome
Spinalis vent. A.
Notochord
Sympathetic gang.
Dors. segmental A.
Dors. aorta
Vagus 10 N.
Aortic arch 6
Ext. jugular V.
Truncus arteriosus
Pericardial cavity
Hyomandibular, branchial groove 1
Ectoderm
Facial V.
Pigment layer
Lens
Cornea
Sensory layer
Ant. cerebral Aa.
Marginal layer
Ependymal layer
Amniotic cavity

261

10 mm. pig embryo
x 40, x. sec., section 24

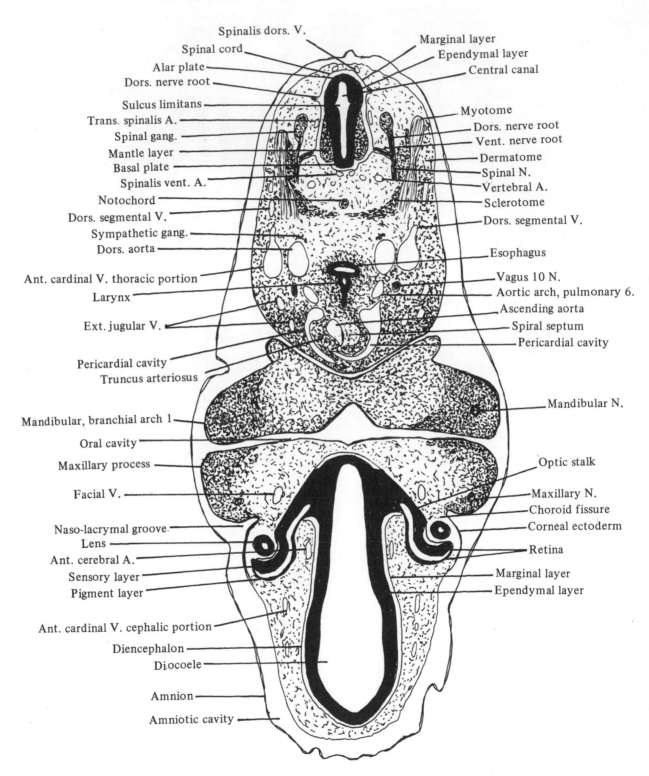

Spinalis dors. V.
Spinal cord
Alar plate
Dors. nerve root
Sulcus limitans
Trans. spinalis A.
Spinal gang.
Mantle layer
Basal plate
Spinalis vent. A.
Notochord
Dors. segmental V.
Sympathetic gang.
Dors. aorta
Ant. cardinal V. thoracic portion
Larynx
Ext. jugular V.
Pericardial cavity
Truncus arteriosus
Mandibular, branchial arch 1
Oral cavity
Maxillary process
Facial V.
Naso-lacrymal groove
Lens
Ant. cerebral A.
Sensory layer
Pigment layer
Ant. cardinal V. cephalic portion
Diencephalon
Diocoele
Amnion
Amniotic cavity

Marginal layer
Ependymal layer
Central canal
Myotome
Dors. nerve root
Vent. nerve root
Dermatome
Spinal N.
Vertebral A.
Sclerotome
Dors. segmental V.
Esophagus
Vagus 10 N.
Aortic arch, pulmonary 6.
Ascending aorta
Spiral septum
Pericardial cavity
Mandibular N.
Optic stalk
Maxillary N.
Choroid fissure
Corneal ectoderm
Retina
Marginal layer
Ependymal layer

10 mm. pig embryo
x 40, x. sec., section 25

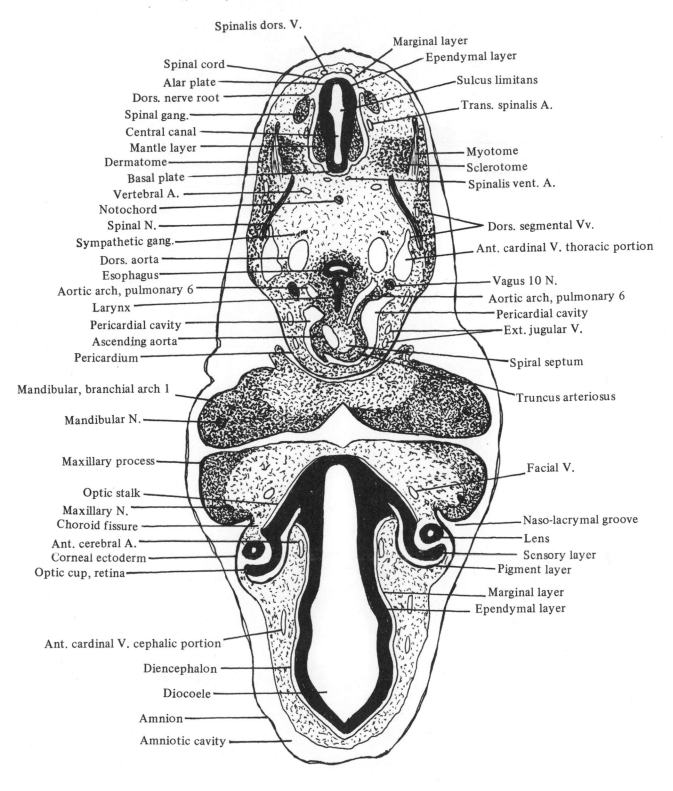

Spinalis dors. V.

Spinal cord

Alar plate

Dors. nerve root

Spinal gang.

Central canal

Mantle layer

Dermatome

Basal plate

Vertebral A.

Notochord

Spinal N.

Sympathetic gang.

Dors. aorta

Esophagus

Aortic arch, pulmonary 6

Larynx

Pericardial cavity

Ascending aorta

Pericardium

Mandibular, branchial arch 1

Mandibular N.

Maxillary process

Optic stalk

Maxillary N.

Choroid fissure

Ant. cerebral A.

Corneal ectoderm

Optic cup, retina

Ant. cardinal V. cephalic portion

Diencephalon

Diocoele

Amnion

Amniotic cavity

Marginal layer

Ependymal layer

Sulcus limitans

Trans. spinalis A.

Myotome

Sclerotome

Spinalis vent. A.

Dors. segmental Vv.

Ant. cardinal V. thoracic portion

Vagus 10 N.

Aortic arch, pulmonary 6

Pericardial cavity

Ext. jugular V.

Spiral septum

Truncus arteriosus

Facial V.

Naso-lacrymal groove

Lens

Sensory layer

Pigment layer

Marginal layer

Ependymal layer

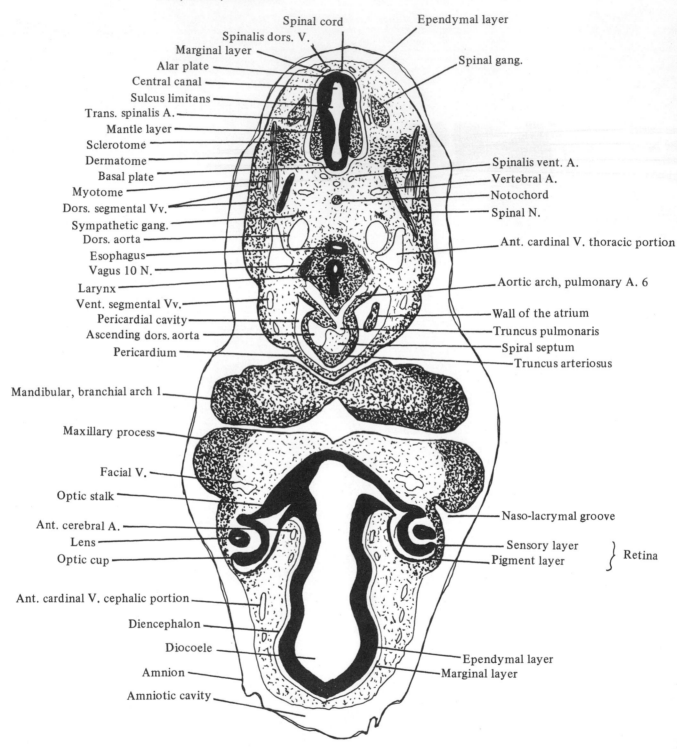

10 mm. pig embryo
x 40, x. sec., section 26

Spinal cord
Spinalis dors. V.
Marginal layer
Alar plate
Central canal
Sulcus limitans
Trans. spinalis A.
Mantle layer
Sclerotome
Dermatome
Basal plate
Myotome
Dors. segmental Vv.
Sympathetic gang.
Dors. aorta
Esophagus
Vagus 10 N.
Larynx
Vent. segmental Vv.
Pericardial cavity
Ascending dors. aorta
Pericardium

Mandibular, branchial arch 1

Maxillary process

Facial V.

Optic stalk

Ant. cerebral A.
Lens
Optic cup

Ant. cardinal V. cephalic portion

Diencephalon

Diocoele

Amnion

Amniotic cavity

Ependymal layer

Spinal gang.

Spinalis vent. A.
Vertebral A.
Notochord
Spinal N.

Ant. cardinal V. thoracic portion

Aortic arch, pulmonary A. 6

Wall of the atrium
Truncus pulmonaris
Spiral septum
Truncus arteriosus

Naso-lacrymal groove

Sensory layer
Pigment layer
} Retina

Ependymal layer
Marginal layer

10 mm. pig embryo
x 40, x. sec., section 27

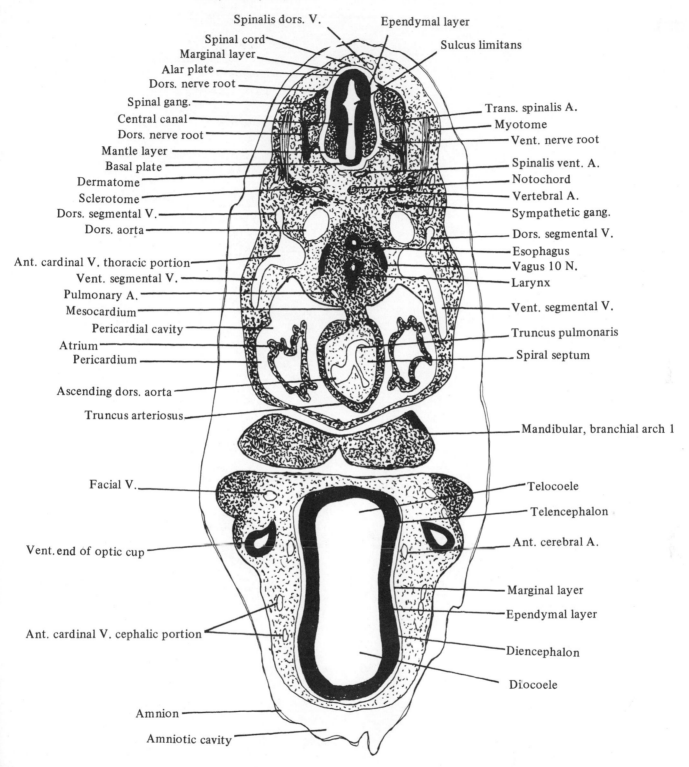

Spinalis dors. V.
Spinal cord
Marginal layer
Alar plate
Dors. nerve root
Spinal gang.
Central canal
Dors. nerve root
Mantle layer
Basal plate
Dermatome
Sclerotome
Dors. segmental V.
Dors. aorta
Ant. cardinal V. thoracic portion
Vent. segmental V.
Pulmonary A.
Mesocardium
Pericardial cavity
Atrium
Pericardium
Ascending dors. aorta
Truncus arteriosus

Facial V.

Vent. end of optic cup

Ant. cardinal V. cephalic portion

Amnion
Amniotic cavity

Ependymal layer
Sulcus limitans

Trans. spinalis A.
Myotome
Vent. nerve root
Spinalis vent. A.
Notochord
Vertebral A.
Sympathetic gang.
Dors. segmental V.
Esophagus
Vagus 10 N.
Larynx
Vent. segmental V.
Truncus pulmonaris
Spiral septum

Mandibular, branchial arch 1

Telocoele
Telencephalon
Ant. cerebral A.

Marginal layer
Ependymal layer

Diencephalon

Diocoele

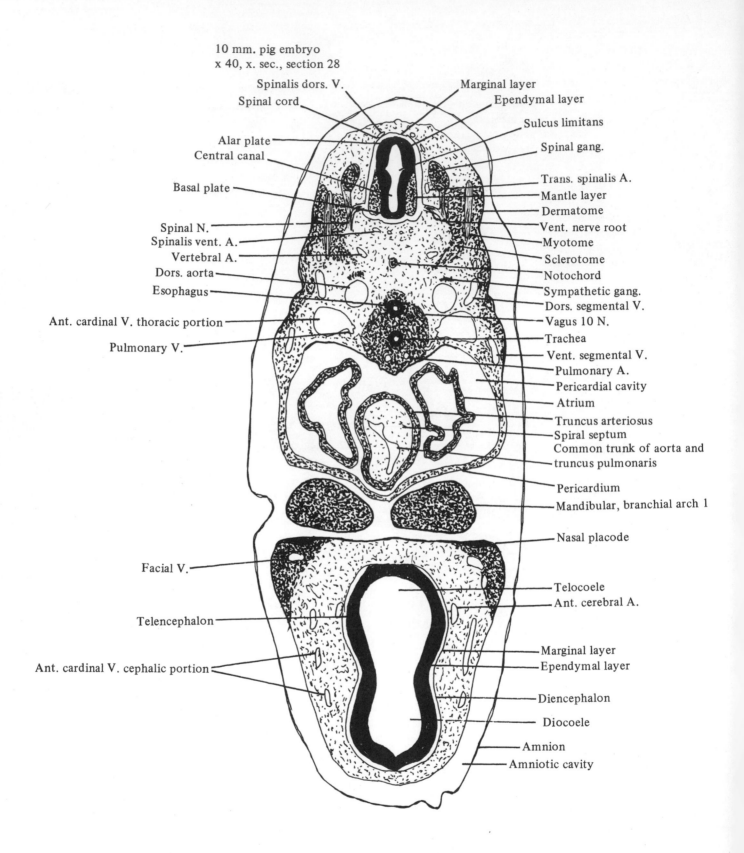

10 mm. pig embryo
x 40, x. sec., section 28

Spinalis dors. V.
Spinal cord
Alar plate
Central canal
Basal plate
Spinal N.
Spinalis vent. A.
Vertebral A.
Dors. aorta
Esophagus
Ant. cardinal V. thoracic portion
Pulmonary V.

Marginal layer
Ependymal layer
Sulcus limitans
Spinal gang.
Trans. spinalis A.
Mantle layer
Dermatome
Vent. nerve root
Myotome
Sclerotome
Notochord
Sympathetic gang.
Dors. segmental V.
Vagus 10 N.
Trachea
Vent. segmental V.
Pulmonary A.
Pericardial cavity
Atrium
Truncus arteriosus
Spiral septum
Common trunk of aorta and
truncus pulmonaris
Pericardium
Mandibular, branchial arch 1
Nasal placode
Telocoele
Ant. cerebral A.
Marginal layer
Ependymal layer
Diencephalon
Diocoele
Amnion
Amniotic cavity

Facial V.
Telencephalon
Ant. cardinal V. cephalic portion

266

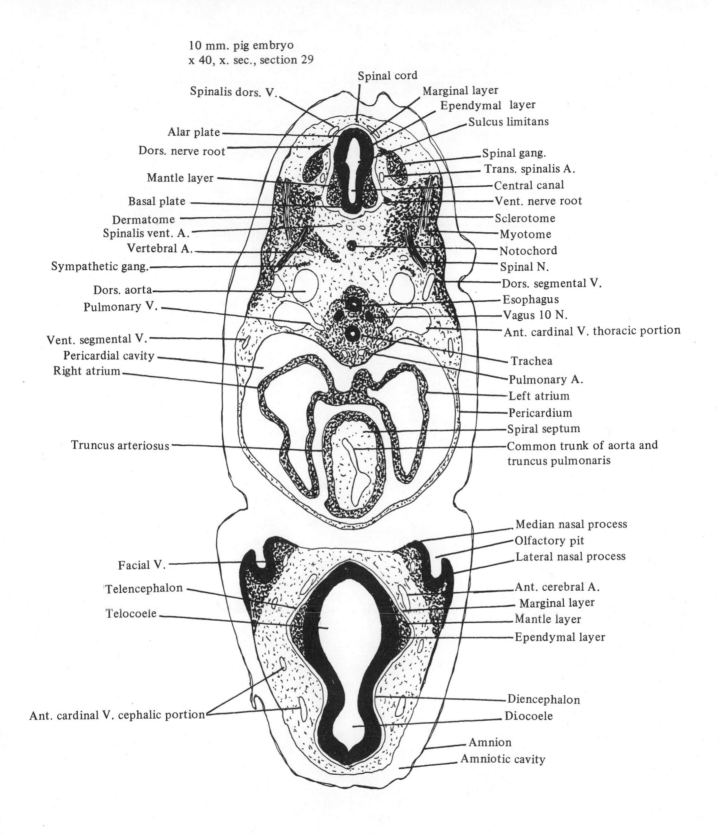

10 mm. pig embryo
x 40, x. sec., section 29

Spinal cord

Spinalis dors. V.

Marginal layer
Ependymal layer
Sulcus limitans

Alar plate
Dors. nerve root

Spinal gang.
Trans. spinalis A.
Central canal
Vent. nerve root

Mantle layer

Basal plate

Dermatome
Spinalis vent. A.
Vertebral A.

Sclerotome
Myotome
Notochord

Sympathetic gang.

Spinal N.

Dors. aorta
Pulmonary V.

Dors. segmental V.
Esophagus
Vagus 10 N.
Ant. cardinal V. thoracic portion

Vent. segmental V.
Pericardial cavity
Right atrium

Trachea
Pulmonary A.
Left atrium
Pericardium
Spiral septum

Truncus arteriosus

Common trunk of aorta and
truncus pulmonaris

Median nasal process
Olfactory pit
Lateral nasal process

Facial V.

Telencephalon

Telocoele

Ant. cerebral A.
Marginal layer
Mantle layer
Ependymal layer

Diencephalon
Diocoele

Ant. cardinal V. cephalic portion

Amnion
Amniotic cavity

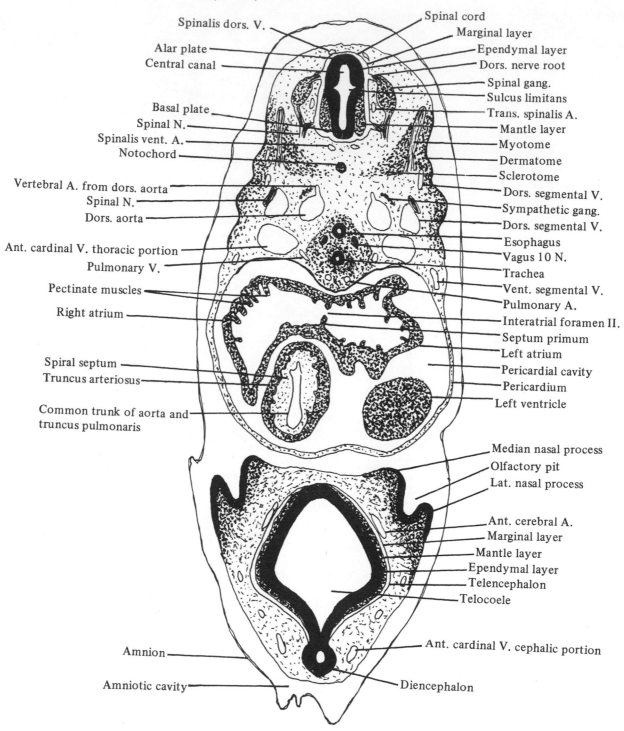

10 mm. pig embryo
x 40, x. sec., section 30

Spinalis dors. V.
Alar plate
Central canal

Spinal cord
Marginal layer
Ependymal layer
Dors. nerve root
Spinal gang.
Sulcus limitans
Trans. spinalis A.
Mantle layer
Myotome
Dermatome
Sclerotome
Dors. segmental V.
Sympathetic gang.
Dors. segmental V.
Esophagus
Vagus 10 N.
Trachea
Vent. segmental V.
Pulmonary A.
Interatrial foramen II.
Septum primum
Left atrium
Pericardial cavity
Pericardium
Left ventricle

Basal plate
Spinal N.
Spinalis vent. A.
Notochord

Vertebral A. from dors. aorta
Spinal N.
Dors. aorta

Ant. cardinal V. thoracic portion
Pulmonary V.

Pectinate muscles
Right atrium

Spiral septum
Truncus arteriosus

Common trunk of aorta and
truncus pulmonaris

Median nasal process
Olfactory pit
Lat. nasal process

Ant. cerebral A.
Marginal layer
Mantle layer
Ependymal layer
Telencephalon
Telocoele

Amnion

Ant. cardinal V. cephalic portion

Amniotic cavity

Diencephalon

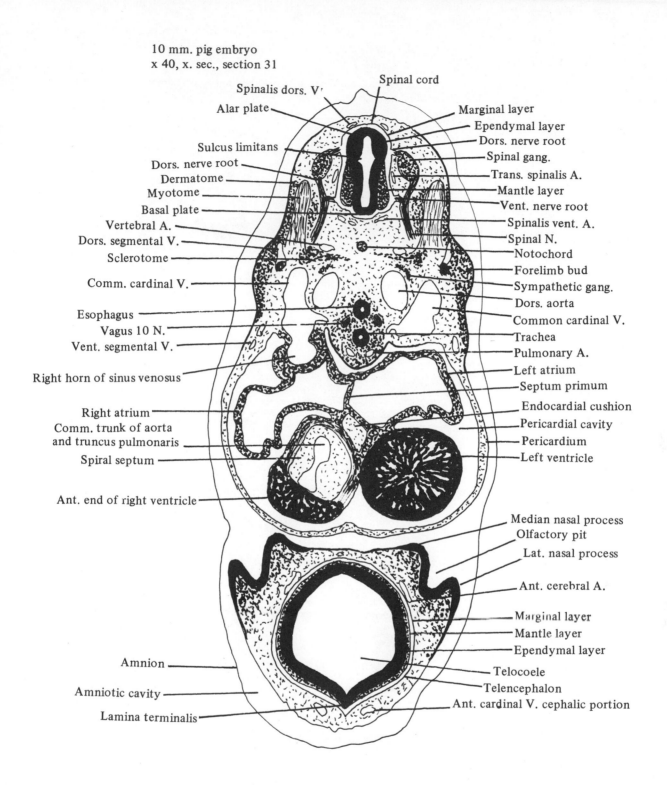

10 mm. pig embryo
x 40, x. sec., section 31

Spinalis dors. V
Spinal cord
Alar plate
Marginal layer
Ependymal layer
Dors. nerve root
Spinal gang.
Sulcus limitans
Dors. nerve root
Dermatome
Myotome
Basal plate
Vertebral A.
Dors. segmental V.
Sclerotome
Comm. cardinal V.
Esophagus
Vagus 10 N.
Vent. segmental V.
Right horn of sinus venosus
Right atrium
Comm. trunk of aorta
and truncus pulmonaris
Spiral septum
Ant. end of right ventricle

Trans. spinalis A.
Mantle layer
Vent. nerve root
Spinalis vent. A.
Spinal N.
Notochord
Forelimb bud
Sympathetic gang.
Dors. aorta
Common cardinal V.
Trachea
Pulmonary A.
Left atrium
Septum primum
Endocardial cushion
Pericardial cavity
Pericardium
Left ventricle

Median nasal process
Olfactory pit
Lat. nasal process
Ant. cerebral A.
Marginal layer
Mantle layer
Ependymal layer

Amnion
Amniotic cavity
Lamina terminalis

Telocoele
Telencephalon
Ant. cardinal V. cephalic portion

10 mm. pig embryo
x 40, x. sec., section 32

Spinalis dors. V.
Spinal cord
Marginal layer
Ependymal layer
Alar plate
Sulcus limitans
Central canal
Dermatome
Myotome
Spinal gang.
Trans. spinalis A.
Basal plate
Mantle layer
Vertebral A.
Spinalis vent. A.
Forelimb bud
Brachial N.
Notochord
Subclavian V.
Dors. aorta
Sympathetic gang.
Esophagus
Comm. cardinal V.
Trachea
Vagus 10 N.
Pulmonary A.
Pulmonary cavity
R. horn of sinus venosus
Left horn of sinus venosus
Sino-atricular valve
Pulmonary V.
R. atrium
L. atrium
R. auricle
L. ventricle
R. ventricle
Trabeculae carneae
Endocardial cushion
Interventricular foramen
Trabeculae carneae
Pericardium
Amnion
Pericardial cavity
Median nasal process
Amniotic cavity
Lat. nasal process
Ant. cerebral Aa.
Head mesoderm
Marginal layer
Ependymal layer
Telencephalon
Telocoele
Head ectoderm
Ant. cardinal Vv. cephalic portion
Lamina terminalis

270

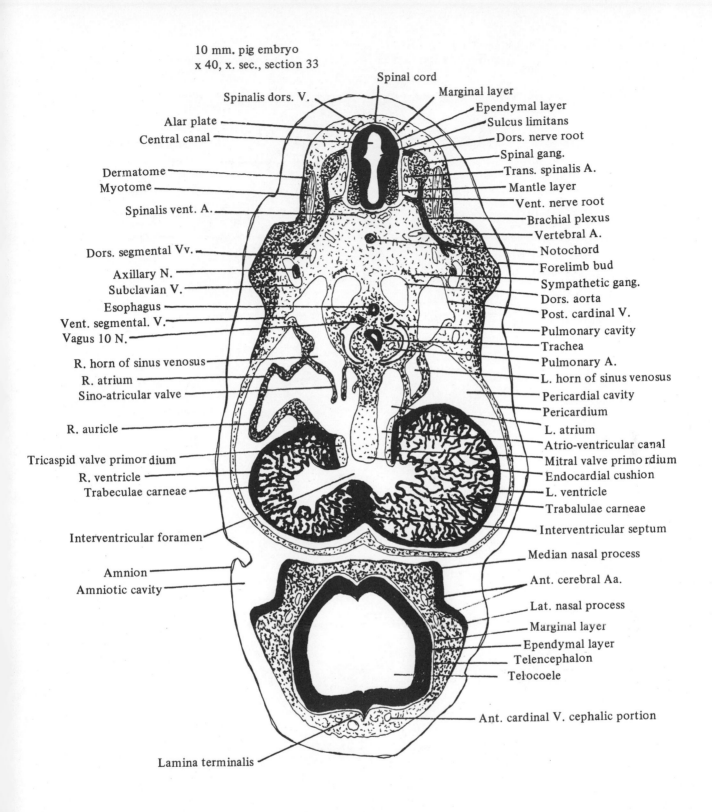

10 mm. pig embryo
x 40, x. sec., section 33

Spinal cord

Spinalis dors. V.

Alar plate

Central canal

Dermatome

Myotome

Spinalis vent. A.

Dors. segmental Vv.

Axillary N.

Subclavian V.

Esophagus

Vent. segmental. V.

Vagus 10 N.

R. horn of sinus venosus

R. atrium

Sino-atricular valve

R. auricle

Tricaspid valve primordium

R. ventricle

Trabeculae carneae

Interventricular foramen

Amnion

Amniotic cavity

Marginal layer

Ependymal layer

Sulcus limitans

Dors. nerve root

Spinal gang.

Trans. spinalis A.

Mantle layer

Vent. nerve root

Brachial plexus

Vertebral A.

Notochord

Forelimb bud

Sympathetic gang.

Dors. aorta

Post. cardinal V.

Pulmonary cavity

Trachea

Pulmonary A.

L. horn of sinus venosus

Pericardial cavity

Pericardium

L. atrium

Atrio-ventricular canal

Mitral valve primordium

Endocardial cushion

L. ventricle

Trabalulae carneae

Interventricular septum

Median nasal process

Ant. cerebral Aa.

Lat. nasal process

Marginal layer

Ependymal layer

Telencephalon

Telocoele

Ant. cardinal V. cephalic portion

Lamina terminalis

271

10 mm. pig embryo
x 40, x. sec., section 34

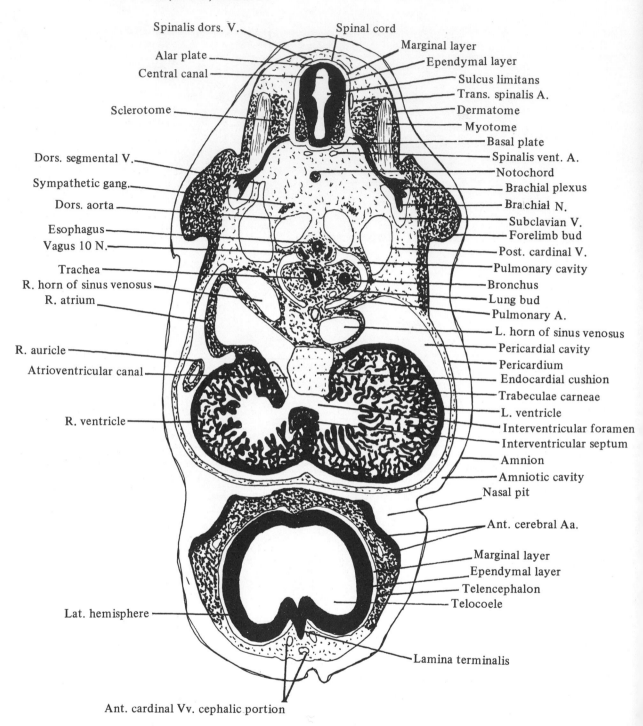

Spinalis dors. V.
Spinal cord
Alar plate
Central canal
Sclerotome
Marginal layer
Ependymal layer
Sulcus limitans
Trans. spinalis A.
Dermatome
Myotome
Basal plate
Spinalis vent. A.
Notochord
Brachial plexus
Brachial N.
Subclavian V.
Forelimb bud
Post. cardinal V.
Pulmonary cavity
Bronchus
Lung bud
Pulmonary A.
L. horn of sinus venosus
Pericardial cavity
Pericardium
Endocardial cushion
Trabeculae carneae
L. ventricle
Interventricular foramen
Interventricular septum
Amnion
Amniotic cavity
Nasal pit
Ant. cerebral Aa.
Marginal layer
Ependymal layer
Telencephalon
Telocoele
Lamina terminalis

Dors. segmental V.
Sympathetic gang.
Dors. aorta
Esophagus
Vagus 10 N.
Trachea
R. horn of sinus venosus
R. atrium
R. auricle
Atrioventricular canal
R. ventricle
Lat. hemisphere
Ant. cardinal Vv. cephalic portion

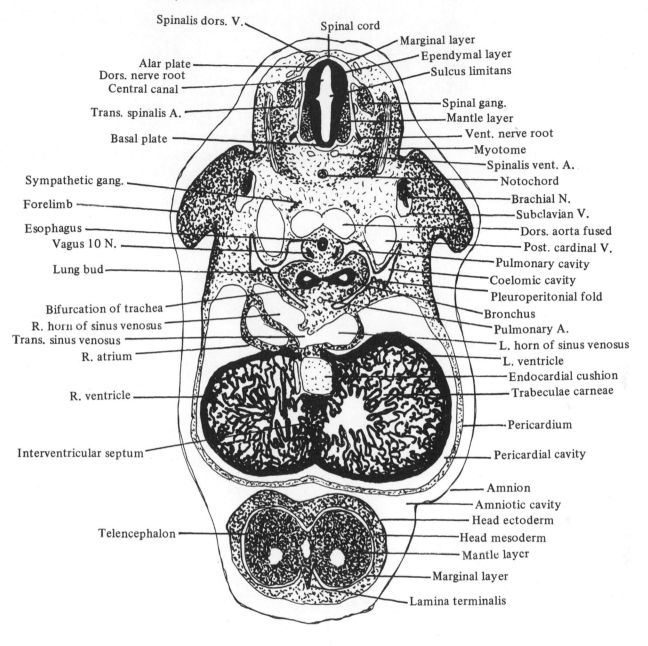

10 mm. pig embryo
x 40, x. sec. section 35

Spinalis dors. V.

Spinal cord

Marginal layer

Ependymal layer

Sulcus limitans

Alar plate

Dors. nerve root

Central canal

Trans. spinalis A.

Basal plate

Spinal gang.

Mantle layer

Vent. nerve root

Myotome

Spinalis vent. A.

Notochord

Sympathetic gang.

Forelimb

Esophagus

Vagus 10 N.

Lung bud

Bifurcation of trachea

R. horn of sinus venosus

Trans. sinus venosus

R. atrium

R. ventricle

Interventricular septum

Brachial N.

Subclavian V.

Dors. aorta fused

Post. cardinal V.

Pulmonary cavity

Coelomic cavity

Pleuroperitonial fold

Bronchus

Pulmonary A.

L. horn of sinus venosus

L. ventricle

Endocardial cushion

Trabeculae carneae

Pericardium

Pericardial cavity

Amnion

Amniotic cavity

Head ectoderm

Head mesoderm

Mantle layer

Marginal layer

Lamina terminalis

Telencephalon

273

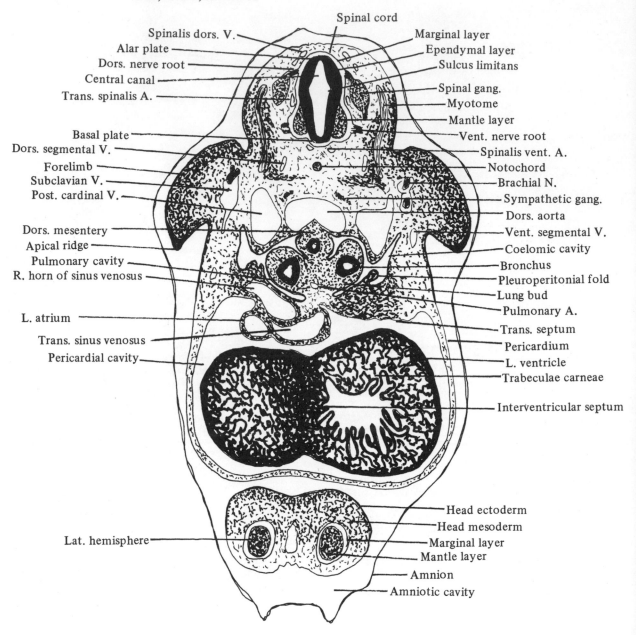

10 mm. pig embryo
x 40, x. sec., section 36

Spinal cord

Spinalis dors. V.
Alar plate
Dors. nerve root
Central canal
Trans. spinalis A.

Marginal layer
Ependymal layer
Sulcus limitans
Spinal gang.
Myotome
Mantle layer
Vent. nerve root
Spinalis vent. A.
Notochord
Brachial N.
Sympathetic gang.
Dors. aorta
Vent. segmental V.
Coelomic cavity
Bronchus
Pleuroperitonial fold
Lung bud
Pulmonary A.
Trans. septum
Pericardium
L. ventricle
Trabeculae carneae

Basal plate
Dors. segmental V.
Forelimb
Subclavian V.
Post. cardinal V.

Dors. mesentery
Apical ridge
Pulmonary cavity
R. horn of sinus venosus

L. atrium

Trans. sinus venosus
Pericardial cavity

Interventricular septum

Head ectoderm
Head mesoderm
Marginal layer
Mantle layer
Amnion
Amniotic cavity

Lat. hemisphere

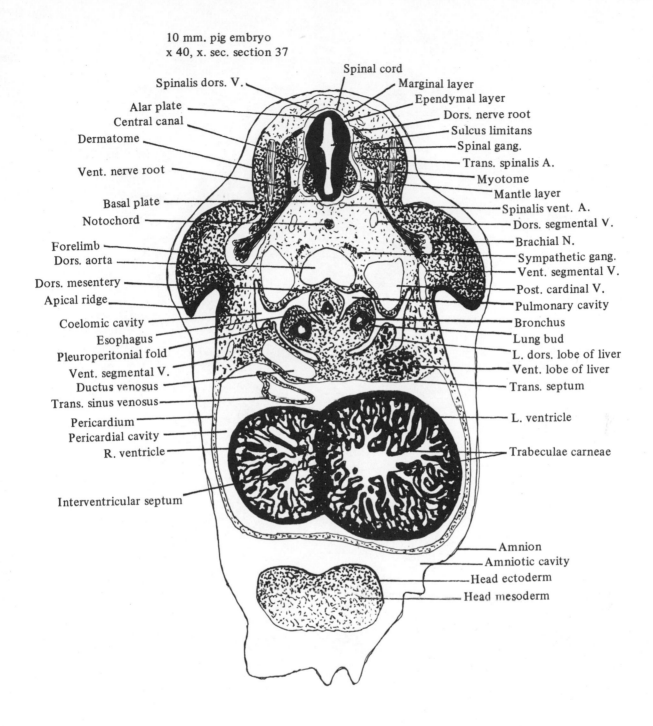

10 mm. pig embryo
x 40, x. sec. section 37

Spinalis dors. V.
Spinal cord
Marginal layer
Ependymal layer
Alar plate
Central canal
Dors. nerve root
Dermatome
Sulcus limitans
Spinal gang.
Vent. nerve root
Trans. spinalis A.
Myotome
Mantle layer
Basal plate
Spinalis vent. A.
Notochord
Dors. segmental V.
Forelimb
Brachial N.
Dors. aorta
Sympathetic gang.
Vent. segmental V.
Dors. mesentery
Post. cardinal V.
Apical ridge
Pulmonary cavity
Coelomic cavity
Bronchus
Esophagus
Lung bud
Pleuroperitonial fold
L. dors. lobe of liver
Vent. segmental V.
Vent. lobe of liver
Ductus venosus
Trans. septum
Trans. sinus venosus
Pericardium
L. ventricle
Pericardial cavity
R. ventricle
Trabeculae carneae

Interventricular septum

Amnion
Amniotic cavity
Head ectoderm
Head mesoderm

275

10 mm. pig embryo
x 40, x. sec., section 38

Alar plate
Spinalis dors. V.
Central canal
Trans. spinalis A.
Vent. nerve root
Basal plate
Spinalis vent. A.
R. subclavian A.
Dors. aorta
R. subclavian V.
Mesonephric tubules
Dors. mesentery
Pulmonary cavity
Coelomic cavity
R. dors. lobe of liver
Omentum bursa
Hepatic cords
Trans. septum
R. ventricle
Interventricular septum

Spinal cord
Marginal layer
Ependymal layer
Dors. nerve root
Sulcus limitans
Spinal gang.
Myotome
Dermatome
Mantle layer
Notochord
Brachial N.
Forelimb
L. subclavian A.
Post. cardinal V.
Vent. segmental V.
Subcardinal V.
Apical ridge
Wall of bronchus
L. dors. lobe of liver
Ductus venosus
Vent. lobe of liver
Pericardial cavity
Pericardium
L. ventricle
Trabeculae carneae

Amnion
Amniotic cavity

10 mm. pig embryo
x 40, x. sec., section 39

Alar plate
Spinalis dors. V.
Central canal
Myotome
Basal plate
Sclerotome
Subclavian V.
Brachial N.
Forelimb
Glomerulus
Subcardinal V.
Vent. segmental V.
Dors. mesogastrum
Caval plica
Omentum bursa
Post. vena cava
Ductus venosus
Hepatic sinus
R. ventricle
Interventricular septum

Spinal cord
Marginal layer
Ependymal layer
Sulcus limitans
Spinal gang.
Trans. spinalis A.
Dermatome
Mantle layer
Spinalis vent. A.
Notochord
Sympathetic gang.
Dors. segmental V.
Dors. aorta
Post. cardinal V.
Mesonephric tubule
Coelomic cavity
Apical ridge
Cardiac portion of stomach
Hepatic sinus
Vent. lobe of liver
Trans. septum
Pericardial cavity
Pericardium
L. ventricle
Trabeculae carneae
Amnion
Amniotic cavity
Ant. edge of umbilicus

277

10 mm. pig embryo
x 40, x. sec., section 40

Alar plate
Spinalis dors. V.
Central canal
Sulcus limitans

Mantle layer
Basal plate
Sympathetic gang.
Dors. segmental V.
Forelimb
Coelomic cavity
Glomerulus
Retiforme central A.

Mesogastrum
Mesonephros
Caval plica
Omentum bursa
Lesser omentum
Post. vena cava
Ductus venosus

Liver
Hepatic cord

R. ventricle

Yolk sac

Spinal cord
Marginal layer
Ependymal layer
Dors. nerve root
Spinal gang.
Trans. spinalis A.
Dermatome
Vent. nerve root
Myotome
Spinalis vent. A.
Notochord
Dors. segmental A.
Brachial N.
Subclavian V.
Post. cardinal V.
Dors. aorta
Mesonephric tubul
Apical ridge
Stomach
L. dors. lobe of live
Hepatic sinuses
Vent. lobe of liver
Amnion
Amniotic cavity
Trans. septum
Pericardial cavity
Pericardium
L. ventricle
Trabeculae carneae

Ant. wall of umbilicus

278

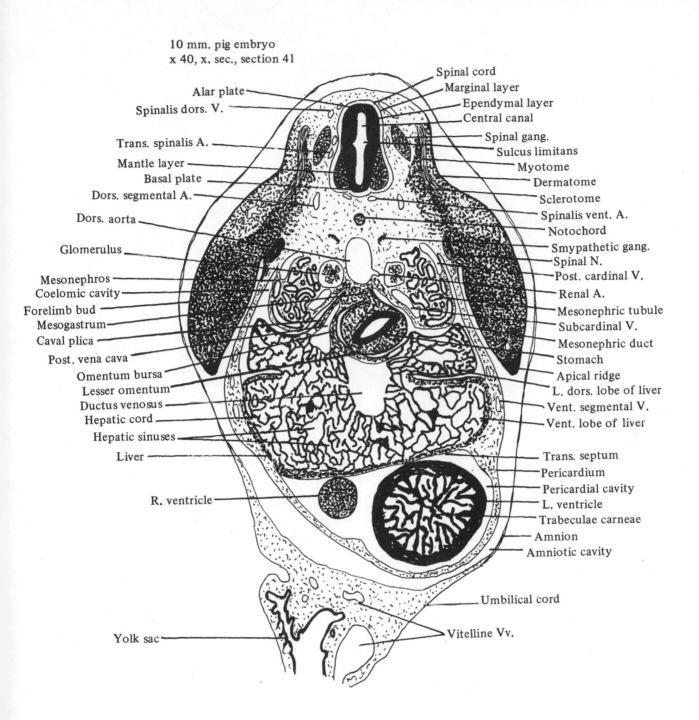

10 mm. pig embryo
x 40, x. sec., section 41

Alar plate
Spinalis dors. V.

Trans. spinalis A.
Mantle layer
Basal plate
Dors. segmental A.
Dors. aorta
Glomerulus
Mesonephros
Coelomic cavity
Forelimb bud
Mesogastrum
Caval plica
Post. vena cava
Omentum bursa
Lesser omentum
Ductus venosus
Hepatic cord
Hepatic sinuses
Liver

R. ventricle

Yolk sac

Spinal cord
Marginal layer
Ependymal layer
Central canal
Spinal gang.
Sulcus limitans
Myotome
Dermatome
Sclerotome
Spinalis vent. A.
Notochord
Smypathetic gang.
Spinal N.
Post. cardinal V.
Renal A.
Mesonephric tubule
Subcardinal V.
Mesonephric duct
Stomach
Apical ridge
L. dors. lobe of liver
Vent. segmental V.
Vent. lobe of liver
Trans. septum
Pericardium
Pericardial cavity
L. ventricle
Trabeculae carneae
Amnion
Amniotic cavity

Umbilical cord

Vitelline Vv.

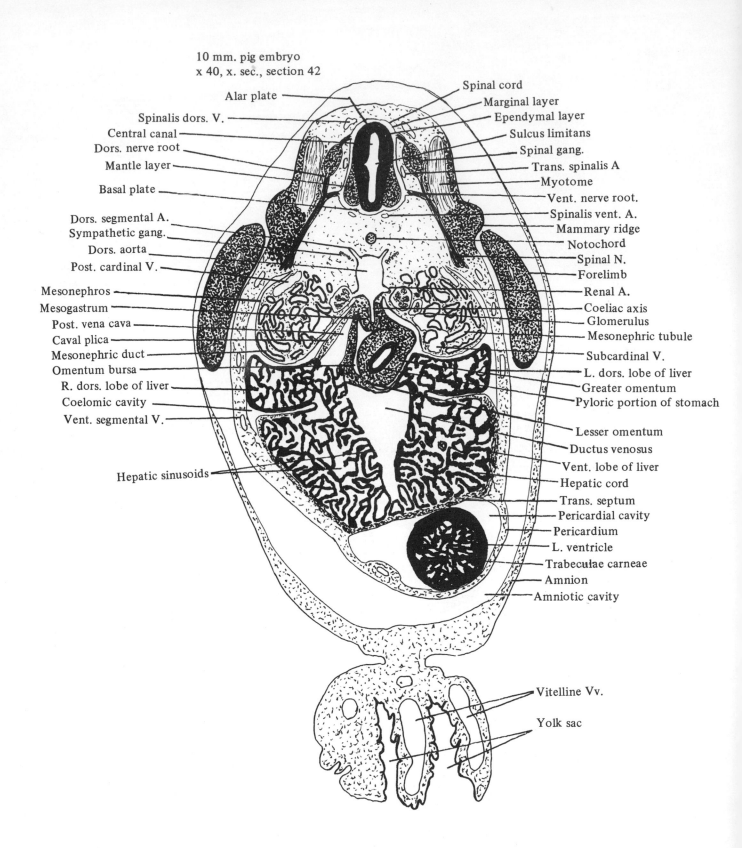

10 mm. pig embryo
x 40, x. sec., section 42

Alar plate
Spinal cord
Spinalis dors. V.
Marginal layer
Central canal
Ependymal layer
Dors. nerve root
Sulcus limitans
Mantle layer
Spinal gang.
Trans. spinalis A
Basal plate
Myotome
Vent. nerve root.
Dors. segmental A.
Spinalis vent. A.
Sympathetic gang.
Mammary ridge
Dors. aorta
Notochord
Post. cardinal V.
Spinal N.
Forelimb
Mesonephros
Renal A.
Mesogastrum
Coeliac axis
Post. vena cava
Glomerulus
Caval plica
Mesonephric tubule
Mesonephric duct
Subcardinal V.
Omentum bursa
L. dors. lobe of liver
R. dors. lobe of liver
Greater omentum
Coelomic cavity
Pyloric portion of stomach
Vent. segmental V.
Lesser omentum
Ductus venosus
Vent. lobe of liver
Hepatic sinusoids
Hepatic cord
Trans. septum
Pericardial cavity
Pericardium
L. ventricle
Trabeculae carneae
Amnion
Amniotic cavity

Vitelline Vv.

Yolk sac

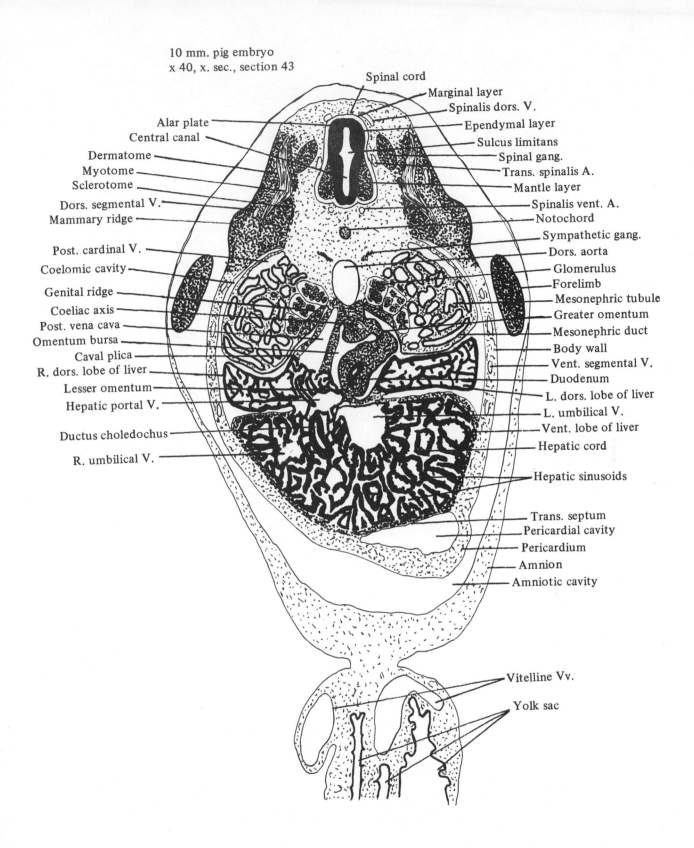

10 mm. pig embryo
x 40, x. sec., section 43

Spinal cord
Marginal layer
Spinalis dors. V.
Ependymal layer
Sulcus limitans
Spinal gang.
Trans. spinalis A.
Mantle layer
Spinalis vent. A.
Notochord
Sympathetic gang.
Dors. aorta
Glomerulus
Forelimb
Mesonephric tubule
Greater omentum
Mesonephric duct
Body wall
Vent. segmental V.
Duodenum
L. dors. lobe of liver
L. umbilical V.
Vent. lobe of liver
Hepatic cord
Hepatic sinusoids
Trans. septum
Pericardial cavity
Pericardium
Amnion
Amniotic cavity

Alar plate
Central canal
Dermatome
Myotome
Sclerotome
Dors. segmental V.
Mammary ridge
Post. cardinal V.
Coelomic cavity
Genital ridge
Coeliac axis
Post. vena cava
Omentum bursa
Caval plica
R. dors. lobe of liver
Lesser omentum
Hepatic portal V.
Ductus choledochus
R. umbilical V.

Vitelline Vv.
Yolk sac

10 mm. pig embryo
x 40, x. sec., section 44

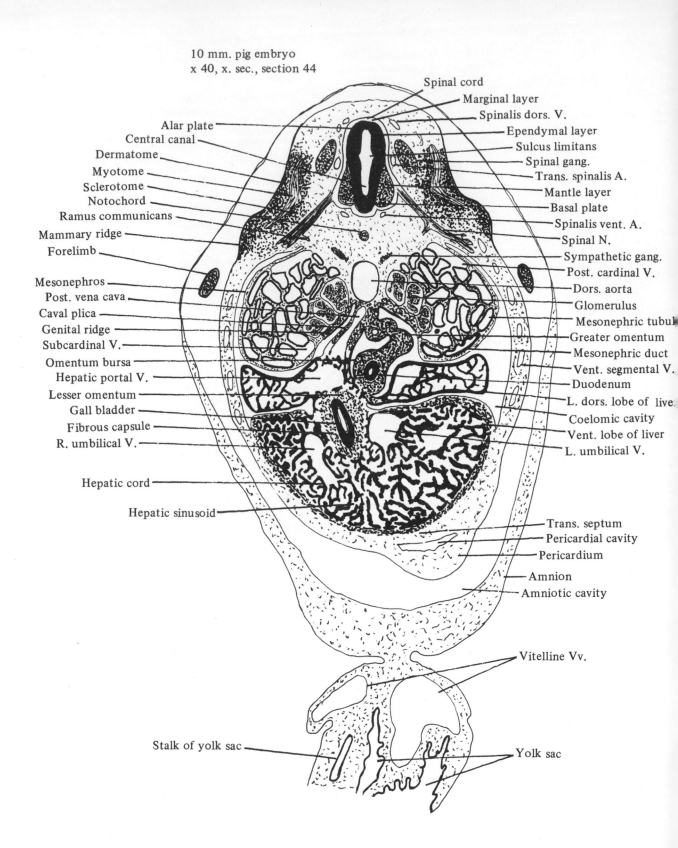

Spinal cord
Marginal layer
Spinalis dors. V.
Ependymal layer
Sulcus limitans
Spinal gang.
Trans. spinalis A.
Mantle layer
Basal plate
Spinalis vent. A.
Spinal N.
Sympathetic gang.
Post. cardinal V.
Dors. aorta
Glomerulus
Mesonephric tubul
Greater omentum
Mesonephric duct
Vent. segmental V.
Duodenum
L. dors. lobe of liver
Coelomic cavity
Vent. lobe of liver
L. umbilical V.

Alar plate
Central canal
Dermatome
Myotome
Sclerotome
Notochord
Ramus communicans
Mammary ridge
Forelimb
Mesonephros
Post. vena cava
Caval plica
Genital ridge
Subcardinal V.
Omentum bursa
Hepatic portal V.
Lesser omentum
Gall bladder
Fibrous capsule
R. umbilical V.

Hepatic cord

Hepatic sinusoid

Trans. septum
Pericardial cavity
Pericardium
Amnion
Amniotic cavity

Vitelline Vv.

Stalk of yolk sac

Yolk sac

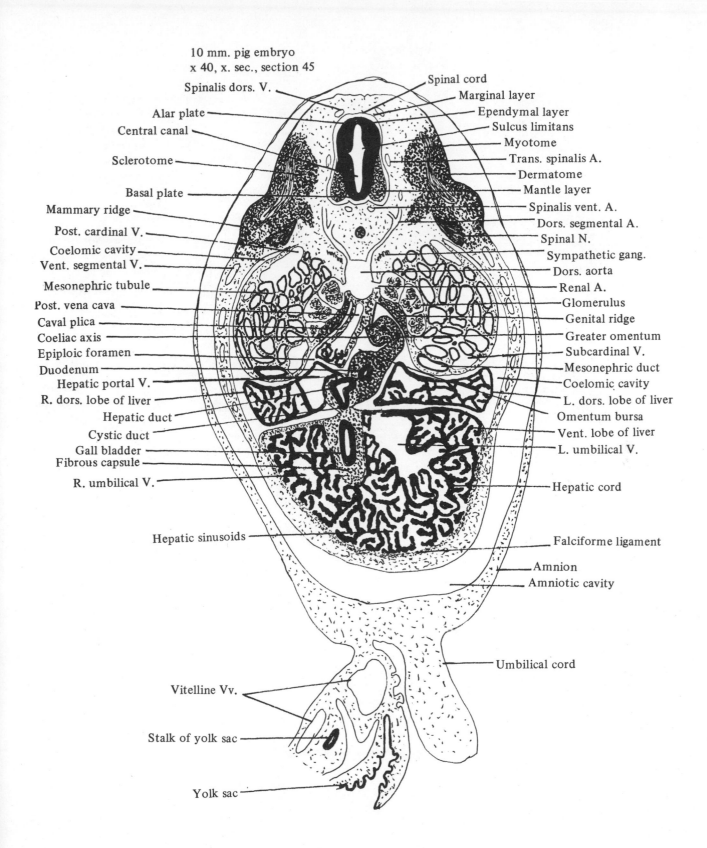

10 mm. pig embryo
x 40, x. sec., section 45

Spinalis dors. V.
Alar plate
Central canal
Sclerotome
Basal plate
Mammary ridge
Post. cardinal V.
Coelomic cavity
Vent. segmental V.
Mesonephric tubule
Post. vena cava
Caval plica
Coeliac axis
Epiploic foramen
Duodenum
Hepatic portal V.
R. dors. lobe of liver
Hepatic duct
Cystic duct
Gall bladder
Fibrous capsule
R. umbilical V.

Hepatic sinusoids

Vitelline Vv.

Stalk of yolk sac

Yolk sac

Spinal cord
Marginal layer
Ependymal layer
Sulcus limitans
Myotome
Trans. spinalis A.
Dermatome
Mantle layer
Spinalis vent. A.
Dors. segmental A.
Spinal N.
Sympathetic gang.
Dors. aorta
Renal A.
Glomerulus
Genital ridge
Greater omentum
Subcardinal V.
Mesonephric duct
Coelomic cavity
L. dors. lobe of liver
Omentum bursa
Vent. lobe of liver
L. umbilical V.

Hepatic cord

Falciforme ligament
Amnion
Amniotic cavity

Umbilical cord

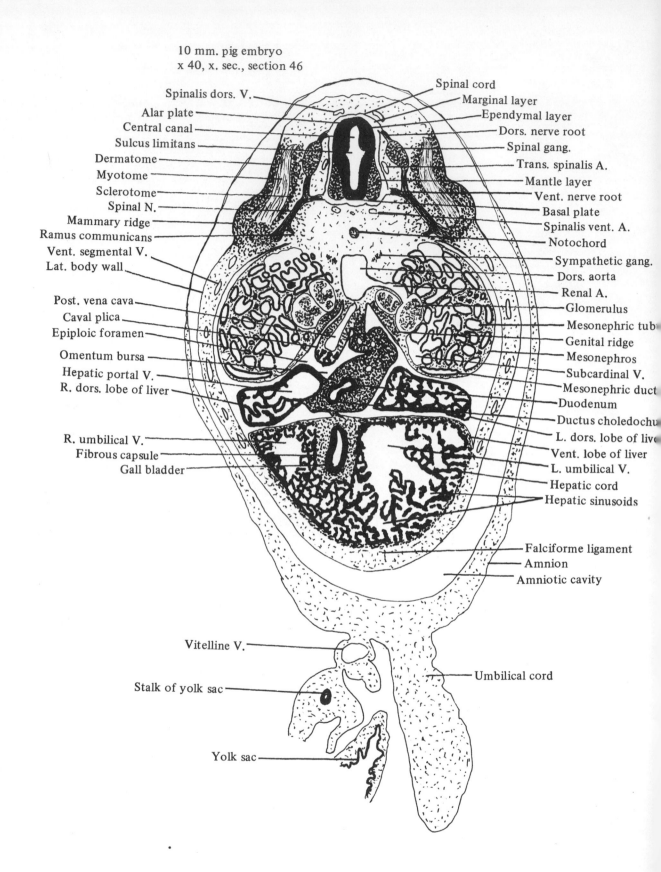

10 mm. pig embryo
x 40, x. sec., section 46

Spinalis dors. V.
Alar plate
Central canal
Sulcus limitans
Dermatome
Myotome
Sclerotome
Spinal N.
Mammary ridge
Ramus communicans
Vent. segmental V.
Lat. body wall

Post. vena cava
Caval plica
Epiploic foramen
Omentum bursa
Hepatic portal V.
R. dors. lobe of liver

R. umbilical V.
Fibrous capsule
Gall bladder

Spinal cord
Marginal layer
Ependymal layer
Dors. nerve root
Spinal gang.
Trans. spinalis A.
Mantle layer
Vent. nerve root
Basal plate
Spinalis vent. A.
Notochord
Sympathetic gang.
Dors. aorta
Renal A.
Glomerulus
Mesonephric tub
Genital ridge
Mesonephros
Subcardinal V.
Mesonephric duct
Duodenum
Ductus choledochu
L. dors. lobe of live
Vent. lobe of liver
L. umbilical V.
Hepatic cord
Hepatic sinusoids

Falciforme ligament
Amnion
Amniotic cavity

Vitelline V.

Stalk of yolk sac

Yolk sac

Umbilical cord

10 mm. pig embryo
x 40, x. sec., section 47

Marginal layer
Ependymal layer
Alar plate
Central canal
Myotome
Dermatome
Sclerotome
Basal plate
Spinal N.

Renal A.
Coelomic cavity
Caval plica
Post. vena cava
Mesonephros
Dors. mesentery
Subcardinal V.
Hepatic portal V.
Ductus choledochus
Lesser omentum
R. dors. lobe of liver

R. umbilical V.

Fibrous capsule

Gall bladder

Spinal cord
Spinalis dors. V.
Dors. nerve root
Sulcus limitans
Spinal gang.
Trans. spinalis A.
Myotome
Vent. nerve root
Mantle layer
Spinalis vent. A.
Mammary ridge
Notochord
Post. cardinal V.
Sympathetic gang.
Dors. aorta
Glomerulus
Lat. body wall
Genital ridge
Mesonephric tubule
Dors. lobe of pancreas
Mesonephric duct
Duodenum
L. dors. lobe of liver
Vent. segmental Vv.
Vent. lobe of liver
Hepatic cord
Hepatic sinusoids
L. umbilical V.
Falciforme ligament
Amnion
Amniotic cavity

Vitelline V.

Stalk of yolk sac

Vitelline A.

Yolk sac

Umbilical cord

285

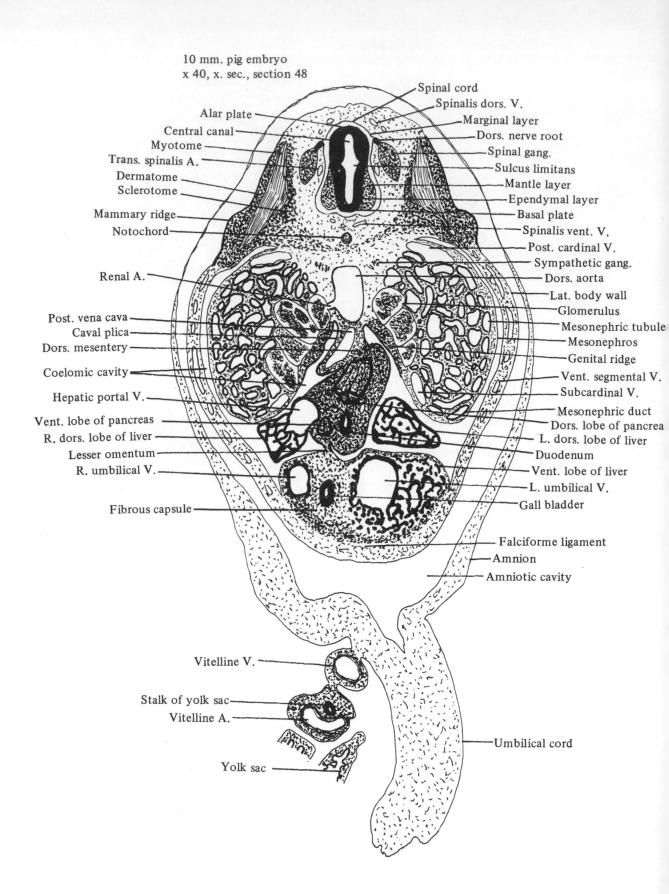

10 mm. pig embryo
x 40, x. sec., section 48

Spinal cord
Spinalis dors. V.
Alar plate
Marginal layer
Central canal
Dors. nerve root
Myotome
Spinal gang.
Trans. spinalis A.
Sulcus limitans
Dermatome
Mantle layer
Sclerotome
Ependymal layer
Mammary ridge
Basal plate
Notochord
Spinalis vent. V.
Post. cardinal V.
Sympathetic gang.
Renal A.
Dors. aorta
Lat. body wall
Glomerulus
Post. vena cava
Mesonephric tubule
Caval plica
Mesonephros
Dors. mesentery
Genital ridge
Coelomic cavity
Vent. segmental V.
Subcardinal V.
Hepatic portal V.
Mesonephric duct
Vent. lobe of pancreas
Dors. lobe of pancrea
R. dors. lobe of liver
L. dors. lobe of liver
Lesser omentum
Duodenum
R. umbilical V.
Vent. lobe of liver
L. umbilical V.
Gall bladder
Fibrous capsule
Falciforme ligament
Amnion
Amniotic cavity
Vitelline V.
Stalk of yolk sac
Vitelline A.
Umbilical cord
Yolk sac

10 mm. pig embryo
x 40, x. sec., section 49

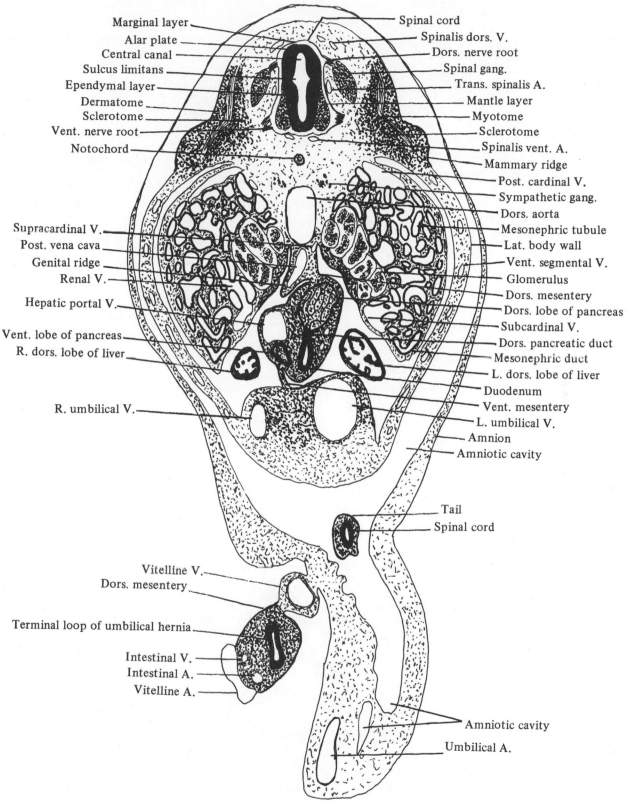

Marginal layer — Spinal cord
Alar plate — Spinalis dors. V.
Central canal — Dors. nerve root
Sulcus limitans — Spinal gang.
Ependymal layer — Trans. spinalis A.
Dermatome — Mantle layer
Sclerotome — Myotome
Vent. nerve root — Sclerotome
Notochord — Spinalis vent. A.
Mammary ridge
Post. cardinal V.
Sympathetic gang.
Supracardinal V. — Dors. aorta
Post. vena cava — Mesonephric tubule
Genital ridge — Lat. body wall
Renal V. — Vent. segmental V.
Glomerulus
Hepatic portal V. — Dors. mesentery
Dors. lobe of pancreas
Subcardinal V.
Vent. lobe of pancreas — Dors. pancreatic duct
R. dors. lobe of liver — Mesonephric duct
L. dors. lobe of liver
Duodenum
Vent. mesentery
R. umbilical V. — L. umbilical V.
Amnion
Amniotic cavity

Tail
Spinal cord

Vitelline V.
Dors. mesentery

Terminal loop of umbilical hernia

Intestinal V.
Intestinal A.
Vitelline A.

Amniotic cavity
Umbilical A.

10 mm. pig embryo
x 40, x. sec, section 50

Spinal cord

Spinalis dors. V.
Alar plate
Central canal
Dermatome
Sclerotome
Vent. nerve root

Basal plate
Spinal N.
Ramus communicans
Post. cardinal V.
Dors. aorta
Renal A.

Post. vena cava

Hepatic portal V.

Vent. lobe of pancreas

Duodenum
Vent. mesentery

Marginal layer
Ependymal layer
Dors. nerve root
Trans. spinalis A.
Sulcus limitans
Mantle layer
Myotome
Spinalis vent. A.
Mammary ridge
Notochord
Sympathetic gang.
Mesonephric tubule
Vent. segmental V.

Mesonephros
Glomerulus
Supracardinal V.
Dors. mesentery
Genital ridge
Dors. lobe of pancreas
Renal V.
Subcardinal V.
Mesonephric duct
Lat. body wall
Coelomic cavity
Anastomosis, L and R. umbilical Vv.

Amnion
Amniotic cavity

Umbilicus
Tail

Tail somite
Spinal cord

Vitelline V.

Small intestine

Intestinal V.

Large intestine
Intestinal A.
Vitelline A.

Allantois

Umbilical A.

288

10 mm. pig embryo
x 40, x. sec., section 51

Spinalis dors. V.
Alar plate
Central canal
Myotome
Sclerotome
Vent. nerve root
Basal plate
Mammary ridge
Notochord
Dors. segmental A.
Vent. segmental V.

Post. vena cava

Small intestine
Vent. mesentery
R. umbilical V.

Vitelline V.

Small intestine
Umbilical hernia
Superior mesenteric V.
Large intestine
Intestinal A.
Vitelline A.

Umbilical V.
Allantoic sac

Spinal cord
Marginal layer
Ependymal layer
Dors. nerve root
Spinal gang.
Trans. spinalis A.
Dermatome
Mantle layer
Spinalis vent. A.
Spinal N.
Sympathetic gang.
Post. cardinal V.
Dors. aorta
Mesonephros
Mesonephric tubule
Renal A.
Glomerulus
Supracardinal V.
Genital ridge
Dors. mesentery
Lat. body wall
Subcardinal V.
Mesonephric duct
Hepatic portal V.
Coelomic cavity
L. umbilical V.

Amnion
Amniotic cavity

Tail
Somite
Notochord

Somites

L. umbilical A.

Spinal cord

Umbilical A.

289

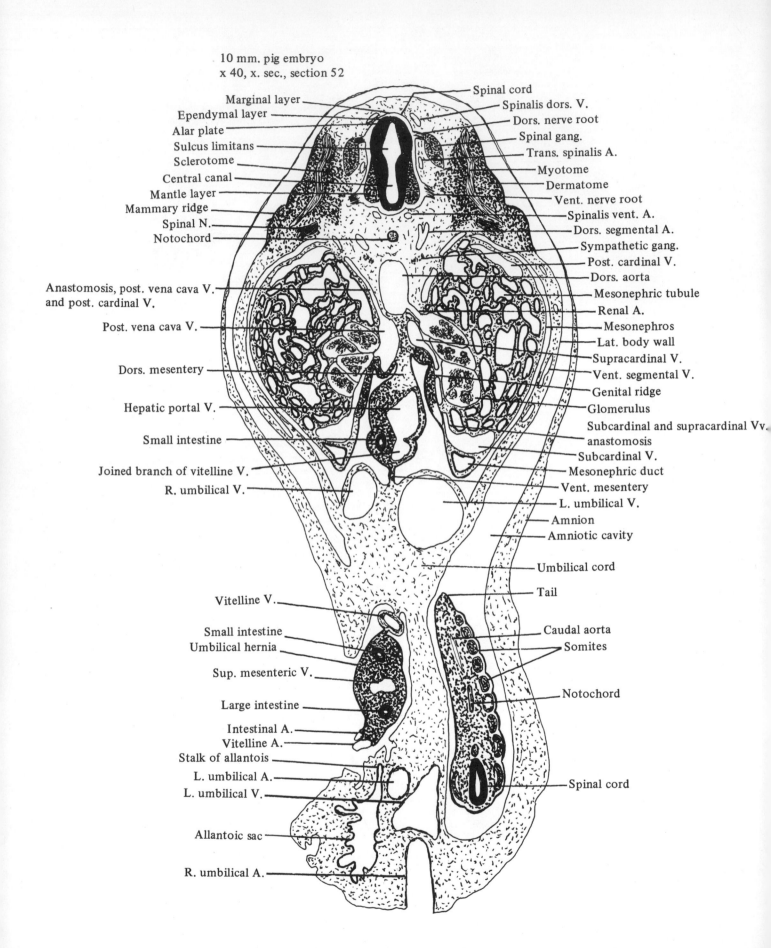

10 mm. pig embryo
x 40, x. sec., section 52

Marginal layer
Ependymal layer
Alar plate
Sulcus limitans
Sclerotome
Central canal
Mantle layer
Mammary ridge
Spinal N.
Notochord

Spinal cord
Spinalis dors. V.
Dors. nerve root
Spinal gang.
Trans. spinalis A.
Myotome
Dermatome
Vent. nerve root
Spinalis vent. A.
Dors. segmental A.
Sympathetic gang.
Post. cardinal V.
Dors. aorta
Mesonephric tubule
Renal A.
Mesonephros
Lat. body wall
Supracardinal V.
Vent. segmental V.
Genital ridge
Glomerulus
Subcardinal and supracardinal Vv.
anastomosis
Subcardinal V.
Mesonephric duct
Vent. mesentery
L. umbilical V.
Amnion
Amniotic cavity

Anastomosis, post. vena cava V.
and post. cardinal V.

Post. vena cava V.

Dors. mesentery

Hepatic portal V.

Small intestine

Joined branch of vitelline V.
R. umbilical V.

Umbilical cord
Tail

Vitelline V.

Small intestine
Umbilical hernia

Sup. mesenteric V.

Large intestine

Intestinal A.
Vitelline A.
Stalk of allantois
L. umbilical A.
L. umbilical V.

Allantoic sac

R. umbilical A.

Caudal aorta
Somites

Notochord

Spinal cord

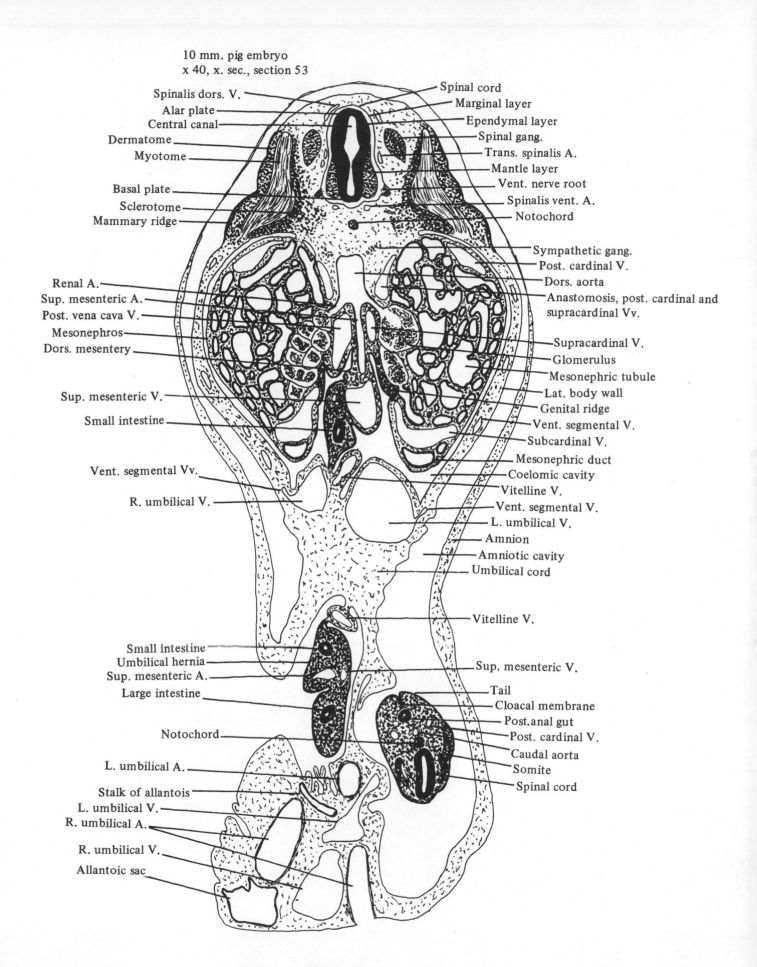

10 mm. pig embryo
x 40, x. sec., section 53

Spinalis dors. V.
Alar plate
Central canal
Dermatome
Myotome

Basal plate
Sclerotome
Mammary ridge

Spinal cord
Marginal layer
Ependymal layer
Spinal gang.
Trans. spinalis A.
Mantle layer
Vent. nerve root
Spinalis vent. A.
Notochord

Renal A.
Sup. mesenteric A.
Post. vena cava V.
Mesonephros
Dors. mesentery

Sympathetic gang.
Post. cardinal V.
Dors. aorta
Anastomosis, post. cardinal and
supracardinal Vv.

Sup. mesenteric V.
Small intestine

Supracardinal V.
Glomerulus
Mesonephric tubule
Lat. body wall
Genital ridge
Vent. segmental V.
Subcardinal V.

Vent. segmental Vv.
R. umbilical V.

Mesonephric duct
Coelomic cavity
Vitelline V.
Vent. segmental V.
L. umbilical V.
Amnion
Amniotic cavity
Umbilical cord

Vitelline V.

Small intestine
Umbilical hernia
Sup. mesenteric A.
Large intestine

Notochord

Sup. mesenteric V.

Tail
Cloacal membrane
Post. anal gut
Post. cardinal V.
Caudal aorta
Somite
Spinal cord

L. umbilical A.
Stalk of allantois
L. umbilical V.
R. umbilical A.
R. umbilical V.
Allantoic sac

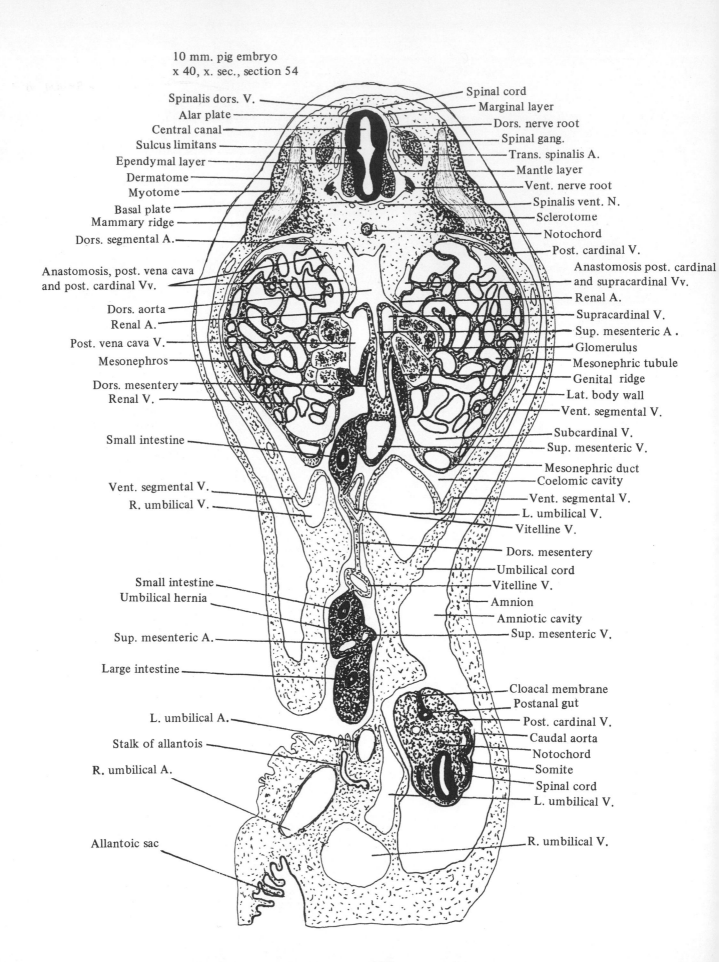

10 mm. pig embryo
x 40, x. sec., section 54

Spinalis dors. V.
Alar plate
Central canal
Sulcus limitans
Ependymal layer
Dermatome
Myotome
Basal plate
Mammary ridge
Dors. segmental A.

Anastomosis, post. vena cava
and post. cardinal Vv.

Dors. aorta
Renal A.
Post. vena cava V.
Mesonephros
Dors. mesentery
Renal V.

Small intestine

Vent. segmental V.
R. umbilical V.

Small intestine
Umbilical hernia

Sup. mesenteric A.

Large intestine

L. umbilical A.

Stalk of allantois

R. umbilical A.

Allantoic sac

Spinal cord
Marginal layer
Dors. nerve root
Spinal gang.
Trans. spinalis A.
Mantle layer
Vent. nerve root
Spinalis vent. N.
Sclerotome
Notochord
Post. cardinal V.

Anastomosis post. cardinal
and supracardinal Vv.

Renal A.
Supracardinal V.
Sup. mesenteric A.
Glomerulus
Mesonephric tubule
Genital ridge
Lat. body wall
Vent. segmental V.
Subcardinal V.
Sup. mesenteric V.

Mesonephric duct
Coelomic cavity
Vent. segmental V.
L. umbilical V.
Vitelline V.

Dors. mesentery
Umbilical cord
Vitelline V.
Amnion
Amniotic cavity
Sup. mesenteric V.

Cloacal membrane
Postanal gut
Post. cardinal V.
Caudal aorta
Notochord
Somite
Spinal cord
L. umbilical V.

R. umbilical V.

292

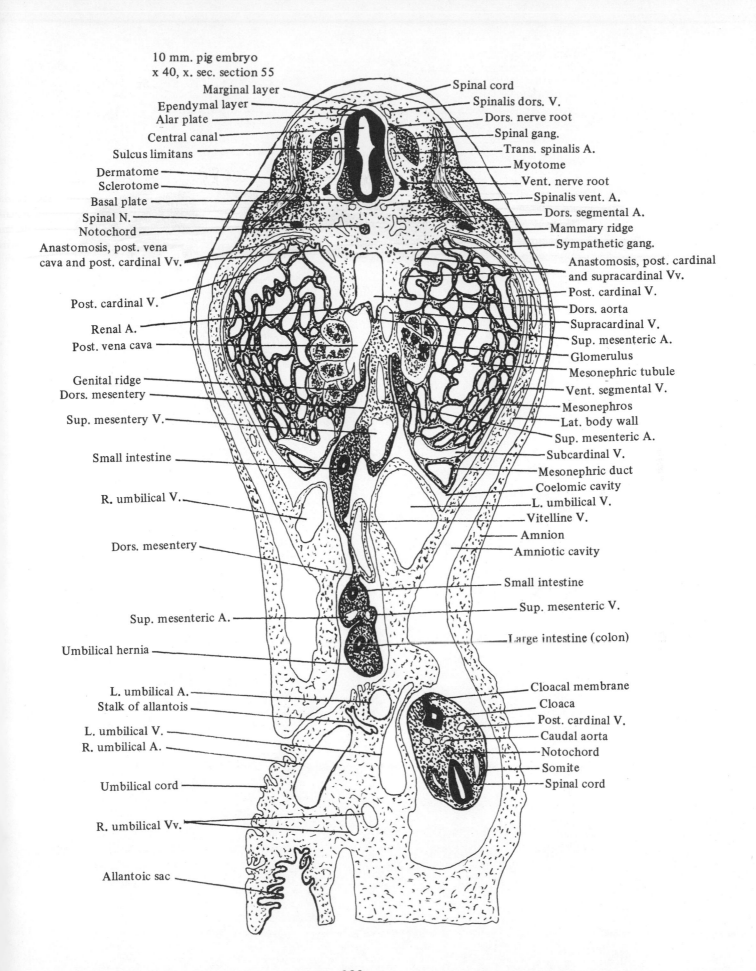

10 mm. pig embryo
x 40, x. sec. section 55

Marginal layer
Ependymal layer
Alar plate
Central canal
Sulcus limitans
Dermatome
Sclerotome
Basal plate
Spinal N.
Notochord
Anastomosis, post. vena
cava and post. cardinal Vv.
Post. cardinal V.
Renal A.
Post. vena cava
Genital ridge
Dors. mesentery
Sup. mesentery V.
Small intestine
R. umbilical V.
Dors. mesentery
Sup. mesenteric A.
Umbilical hernia
L. umbilical A.
Stalk of allantois
L. umbilical V.
R. umbilical A.
Umbilical cord
R. umbilical Vv.
Allantoic sac

Spinal cord
Spinalis dors. V.
Dors. nerve root
Spinal gang.
Trans. spinalis A.
Myotome
Vent. nerve root
Spinalis vent. A.
Dors. segmental A.
Mammary ridge
Sympathetic gang.
Anastomosis, post. cardinal
and supracardinal Vv.
Post. cardinal V.
Dors. aorta
Supracardinal V.
Sup. mesenteric A.
Glomerulus
Mesonephric tubule
Vent. segmental V.
Mesonephros
Lat. body wall
Sup. mesenteric A.
Subcardinal V.
Mesonephric duct
Coelomic cavity
L. umbilical V.
Vitelline V.
Amnion
Amniotic cavity
Small intestine
Sup. mesenteric V.
Large intestine (colon)
Cloacal membrane
Cloaca
Post. cardinal V.
Caudal aorta
Notochord
Somite
Spinal cord

293

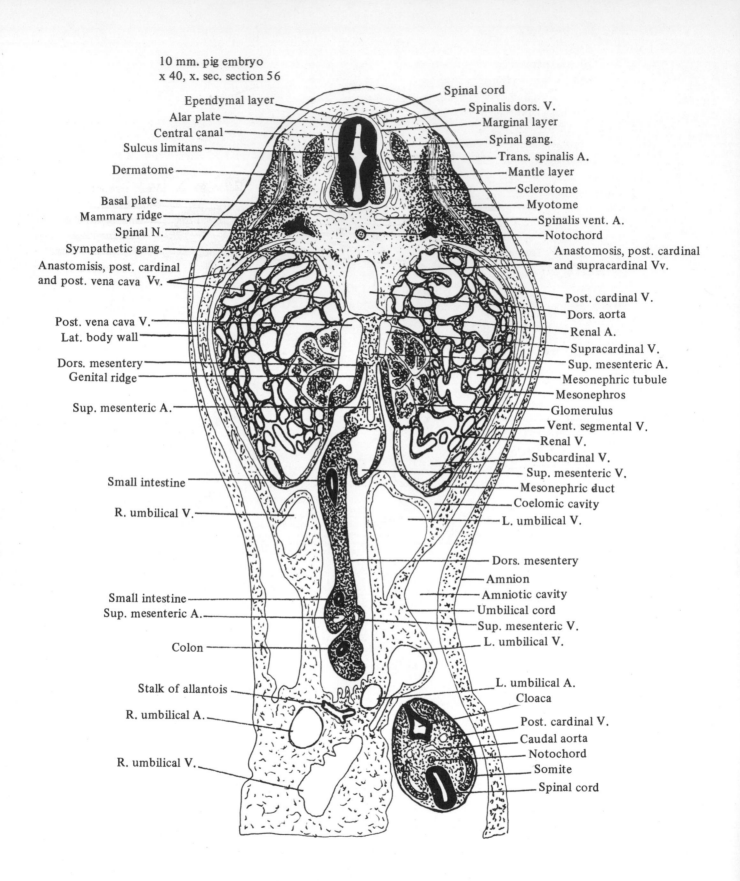

10 mm. pig embryo
x 40, x. sec. section 56

Ependymal layer
Alar plate
Central canal
Sulcus limitans
Dermatome

Basal plate
Mammary ridge
Spinal N.
Sympathetic gang.
Anastomisis, post. cardinal
and post. vena cava Vv.

Post. vena cava V.
Lat. body wall

Dors. mesentery
Genital ridge

Sup. mesenteric A.

Small intestine

R. umbilical V.

Small intestine
Sup. mesenteric A.

Colon

Stalk of allantois

R. umbilical A.

R. umbilical V.

Spinal cord
Spinalis dors. V.
Marginal layer
Spinal gang.
Trans. spinalis A.
Mantle layer
Sclerotome
Myotome
Spinalis vent. A.
Notochord
Anastomosis, post. cardinal
and supracardinal Vv.

Post. cardinal V.
Dors. aorta
Renal A.
Supracardinal V.
Sup. mesenteric A.
Mesonephric tubule
Mesonephros
Glomerulus
Vent. segmental V.
Renal V.
Subcardinal V.
Sup. mesenteric V.
Mesonephric duct
Coelomic cavity
L. umbilical V.

Dors. mesentery
Amnion
Amniotic cavity
Umbilical cord
Sup. mesenteric V.
L. umbilical V.

L. umbilical A.
Cloaca
Post. cardinal V.
Caudal aorta
Notochord
Somite
Spinal cord

10 mm. pig embryo
x 40, x. sec. section 57

Spinal cord
Spinalis dors. V.
Alar plate
Marginal layer
Central canal
Ependymal layer
Sulcus limitans
Trans. spinalis A.
Mantle layer
Sclerotome
Myotome
Basal plate
Dermatome
Spinal N.
Spinalis vent. A.
Mammary ridge
Notochord
Vent. segmental V.
Sympathetic gang.
Anastomosis, post. cardinal
and post. vena cava Vv.
Anastomosis, post. cardinal
and supracardinal Vv.
Post. cardinal V.
Dors. aorta
Renal A.
Lat. body wall
Supracardinal V.
Post. vena cava V.
Inf. mesenteric A.
Mesonephric tubule
Mesonephros
Glomerulus
Vent. segmental V.
Genital ridge
Dors. mesentery
Sup. mesenteric A.
Renal V.
Subcardinal V.
Mesonephric duct
Sup. mesenteric V.
R. umbilical V.
L. umbilical V. abdominal portion
Small intestine
Sup. mesenteric V.
Sup. mesenteric A.
Colon
L. umbilical V. umbilical portion
Stalk of allantois
L. umbilical A.
R. umbilical A.
Cloaca
Post. cardinal V.
Caudal aorta
R. umbilical Vv.
Notochord
Spinal cord
Somite

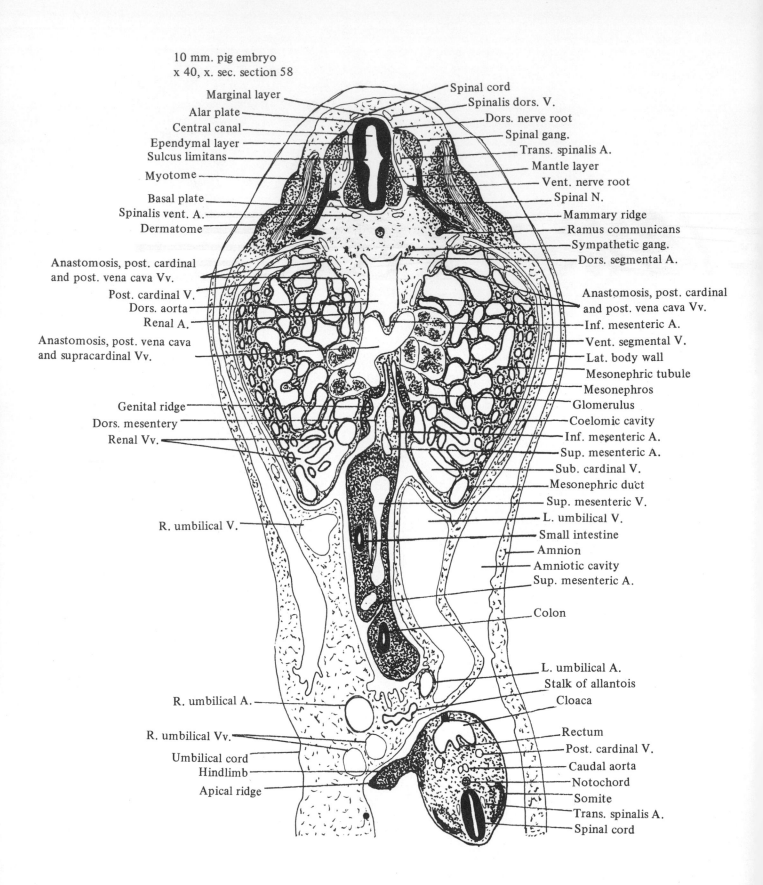

10 mm. pig embryo
x 40, x. sec. section 58

Marginal layer
Alar plate
Central canal
Ependymal layer
Sulcus limitans
Myotome
Basal plate
Spinalis vent. A.
Dermatome

Spinal cord
Spinalis dors. V.
Dors. nerve root
Spinal gang.
Trans. spinalis A.
Mantle layer
Vent. nerve root
Spinal N.
Mammary ridge
Ramus communicans
Sympathetic gang.
Dors. segmental A.

Anastomosis, post. cardinal
and post. vena cava Vv.
Post. cardinal V.
Dors. aorta
Renal A.
Anastomosis, post. vena cava
and supracardinal Vv.

Anastomosis, post. cardinal
and post. vena cava Vv.
Inf. mesenteric A.
Vent. segmental V.
Lat. body wall
Mesonephric tubule
Mesonephros
Glomerulus
Coelomic cavity
Inf. mesenteric A.
Sup. mesenteric A.
Sub. cardinal V.
Mesonephric duct
Sup. mesenteric V.
L. umbilical V.
Small intestine
Amnion
Amniotic cavity
Sup. mesenteric A.

Genital ridge
Dors. mesentery
Renal Vv.

R. umbilical V.

Colon

R. umbilical A.

L. umbilical A.
Stalk of allantois
Cloaca

R. umbilical Vv.
Umbilical cord
Hindlimb
Apical ridge

Rectum
Post. cardinal V.
Caudal aorta
Notochord
Somite
Trans. spinalis A.
Spinal cord

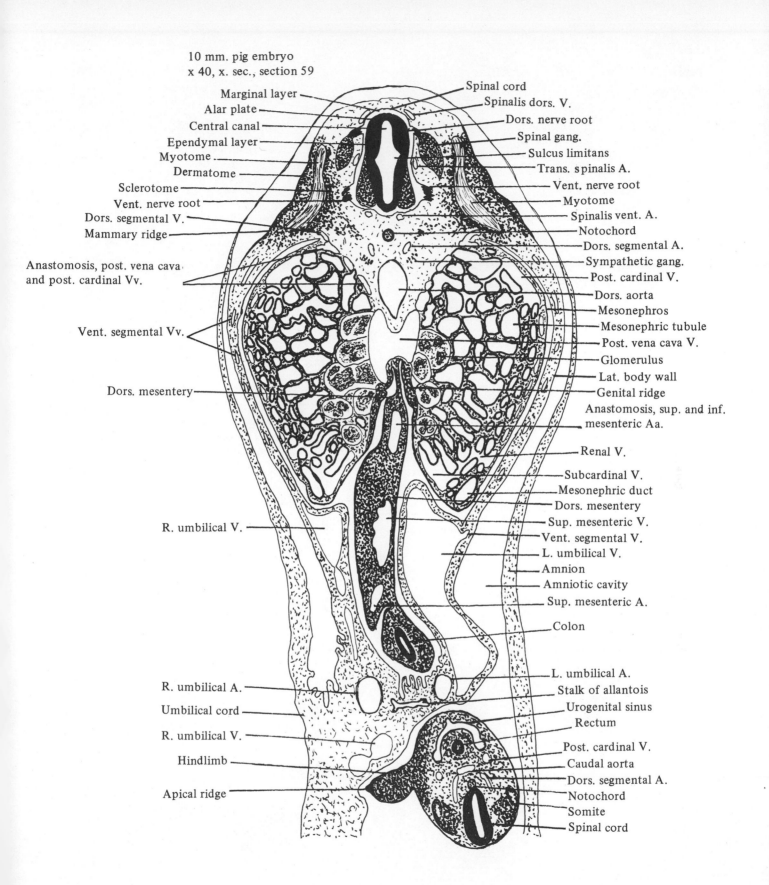

10 mm. pig embryo
x 40, x. sec., section 59

Marginal layer
Alar plate
Central canal
Ependymal layer
Myotome
Dermatome
Sclerotome
Vent. nerve root
Dors. segmental V.
Mammary ridge

Anastomosis, post. vena cava
and post. cardinal Vv.

Vent. segmental Vv.

Dors. mesentery

R. umbilical V.

R. umbilical A.
Umbilical cord
R. umbilical V.
Hindlimb
Apical ridge

Spinal cord
Spinalis dors. V.
Dors. nerve root
Spinal gang.
Sulcus limitans
Trans. spinalis A.
Vent. nerve root
Myotome
Spinalis vent. A.
Notochord
Dors. segmental A.
Sympathetic gang.
Post. cardinal V.
Dors. aorta
Mesonephros
Mesonephric tubule
Post. vena cava V.
Glomerulus
Lat. body wall
Genital ridge
Anastomosis, sup. and inf.
mesenteric Aa.
Renal V.
Subcardinal V.
Mesonephric duct
Dors. mesentery
Sup. mesenteric V.
Vent. segmental V.
L. umbilical V.
Amnion
Amniotic cavity
Sup. mesenteric A.
Colon
L. umbilical A.
Stalk of allantois
Urogenital sinus
Rectum
Post. cardinal V.
Caudal aorta
Dors. segmental A.
Notochord
Somite
Spinal cord

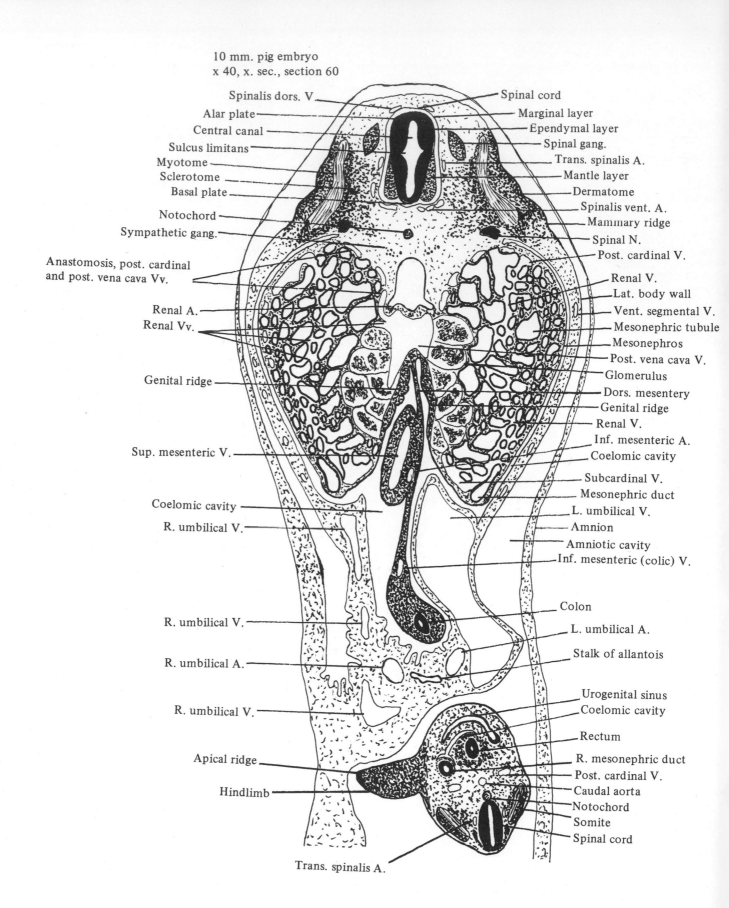

10 mm. pig embryo
x 40, x. sec., section 60

Spinalis dors. V.
Alar plate
Central canal
Sulcus limitans
Myotome
Sclerotome
Basal plate
Notochord
Sympathetic gang.

Anastomosis, post. cardinal
and post. vena cava Vv.

Renal A.
Renal Vv.

Genital ridge

Sup. mesenteric V.

Coelomic cavity
R. umbilical V.

R. umbilical V.

R. umbilical A.

R. umbilical V.

Apical ridge

Hindlimb

Spinal cord
Marginal layer
Ependymal layer
Spinal gang.
Trans. spinalis A.
Mantle layer
Dermatome
Spinalis vent. A.
Mammary ridge
Spinal N.
Post. cardinal V.
Renal V.
Lat. body wall
Vent. segmental V.
Mesonephric tubule
Mesonephros
Post. vena cava V.
Glomerulus
Dors. mesentery
Genital ridge
Renal V.
Inf. mesenteric A.
Coelomic cavity
Subcardinal V.
Mesonephric duct
L. umbilical V.
Amnion
Amniotic cavity
Inf. mesenteric (colic) V.

Colon
L. umbilical A.

Stalk of allantois

Urogenital sinus
Coelomic cavity
Rectum
R. mesonephric duct
Post. cardinal V.
Caudal aorta
Notochord
Somite
Spinal cord

Trans. spinalis A.

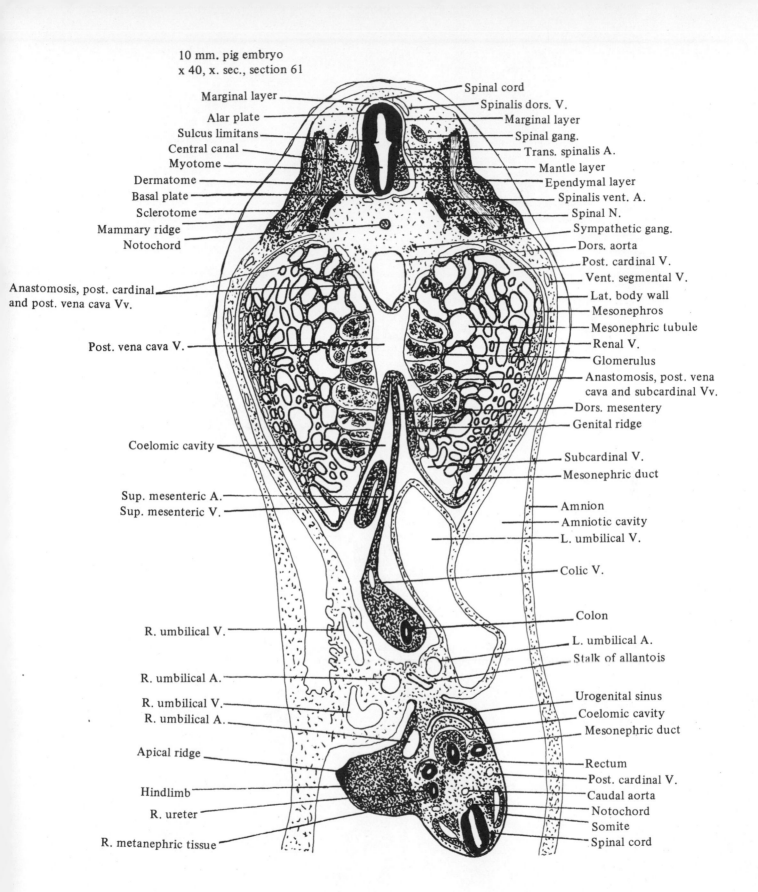

10 mm. pig embryo
x 40, x. sec., section 61

Marginal layer
Alar plate
Sulcus limitans
Central canal
Myotome
Dermatome
Basal plate
Sclerotome
Mammary ridge
Notochord

Spinal cord
Spinalis dors. V.
Marginal layer
Spinal gang.
Trans. spinalis A.
Mantle layer
Ependymal layer
Spinalis vent. A.
Spinal N.
Sympathetic gang.
Dors. aorta
Post. cardinal V.
Vent. segmental V.
Lat. body wall
Mesonephros
Mesonephric tubule
Renal V.
Glomerulus
Anastomosis, post. vena
cava and subcardinal Vv.
Dors. mesentery
Genital ridge

Anastomosis, post. cardinal
and post. vena cava Vv.

Post. vena cava V.

Coelomic cavity

Subcardinal V.
Mesonephric duct

Sup. mesenteric A.
Sup. mesenteric V.

Amnion
Amniotic cavity
L. umbilical V.

Colic V.

R. umbilical V.

Colon

R. umbilical A.

L. umbilical A.
Stalk of allantois

R. umbilical V.
R. umbilical A.

Urogenital sinus
Coelomic cavity
Mesonephric duct

Apical ridge

Rectum
Post. cardinal V.
Caudal aorta
Notochord
Somite
Spinal cord

Hindlimb
R. ureter

R. metanephric tissue

299

10 mm. pig embryo
x 40, x. sec., section 62

Marginal layer
Alar plate
Central canal
Sulcus limitans
Ependymal layer
Dermatome
Myotome
Basal plate
Notochord
Dors. segmental A.

Anastomosis, post. cardinal and
post. vena cava Vv.

Dors. aorta

Post. vena cava V.

Renal Vv.

Vent. segmental V.

R. umbilical V.

R. umbilical A.

Vent. segmental Vv.
R. hindlimb
Apical ridge
Mesonephric duct
Metanephros
Primordial pelvis

Spinal cord
Spinalis dors. V.
Dors. nerve root
Spinal gang.
Trans. spinalis A.
Mantle layer
Vent. nerve root
Spinalis vent. A.
Mammary ridge

Sclerotome
Sympathetic gang.

Post. cardinal V.
Lat. body wall
Mesonephros
Renal V.
Mesonephric tubule
Glomerulus
Anastomosis, post. vena
cava and subcardinal Vv.

Dors. mesentery
Coelomic cavity

Mesonephric duct
Subcardinal V.
Colic A.
Amnion
Amniotic cavity
Vent. segmental V.
L. umbilical V.
Colic V.

Colon

L. umbilical A.
Stalk of allantois enters
into urogenital sinus
L. hindlimb
Coelomic cavity
Rectum
L. mesonephric duct
L. uretric bud
Post. cardinal V.
Caudal aorta
Notochord
Somite
Spinal cord

Trans. spinalis A.

300

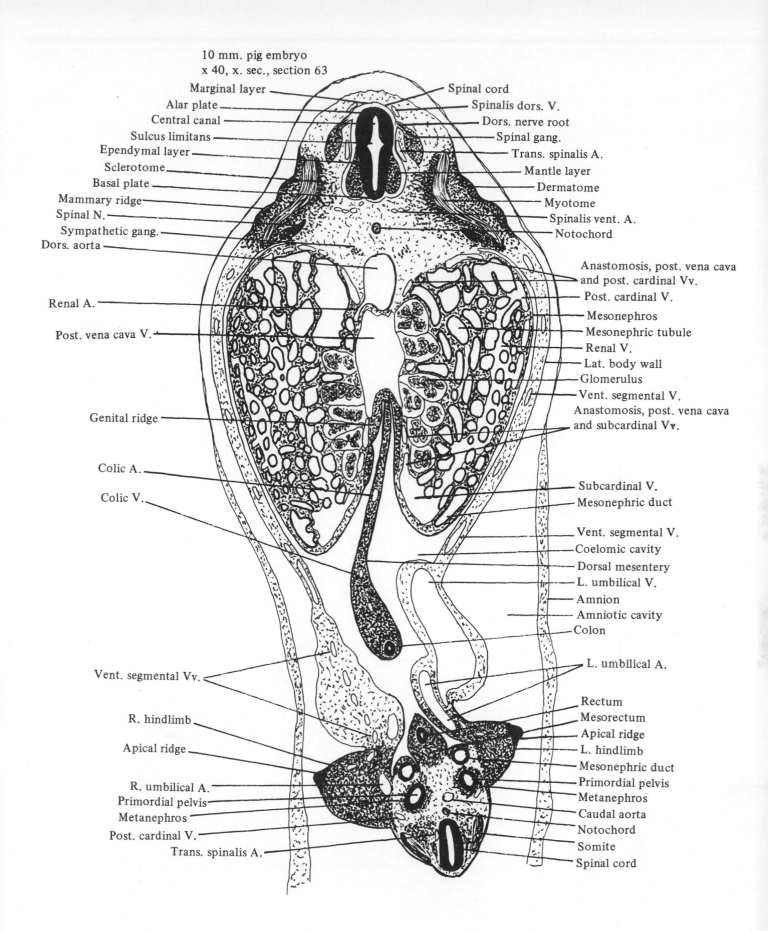

10 mm. pig embryo
x 40, x. sec., section 63

Marginal layer — Spinal cord
Alar plate — Spinalis dors. V.
Central canal — Dors. nerve root
Sulcus limitans — Spinal gang.
Ependymal layer — Trans. spinalis A.
Sclerotome — Mantle layer
Basal plate — Dermatome
Mammary ridge — Myotome
Spinal N. — Spinalis vent. A.
Sympathetic gang. — Notochord
Dors. aorta —

Anastomosis, post. vena cava and post. cardinal Vv.
Renal A. — Post. cardinal V.
Mesonephros
Post. vena cava V. — Mesonephric tubule
Renal V.
Lat. body wall
Glomerulus
Genital ridge — Vent. segmental V.
Anastomosis, post. vena cava and subcardinal Vv.

Colic A. — Subcardinal V.
Colic V. — Mesonephric duct

Vent. segmental V.
Coelomic cavity
Dorsal mesentery
L. umbilical V.
Amnion
Amniotic cavity
Colon

Vent. segmental Vv. — L. umbilical A.

Rectum
Mesorectum
R. hindlimb — Apical ridge
Apical ridge — L. hindlimb
Mesonephric duct
R. umbilical A. — Primordial pelvis
Primordial pelvis — Metanephros
Metanephros — Caudal aorta
Post. cardinal V. — Notochord
Trans. spinalis A. — Somite
Spinal cord

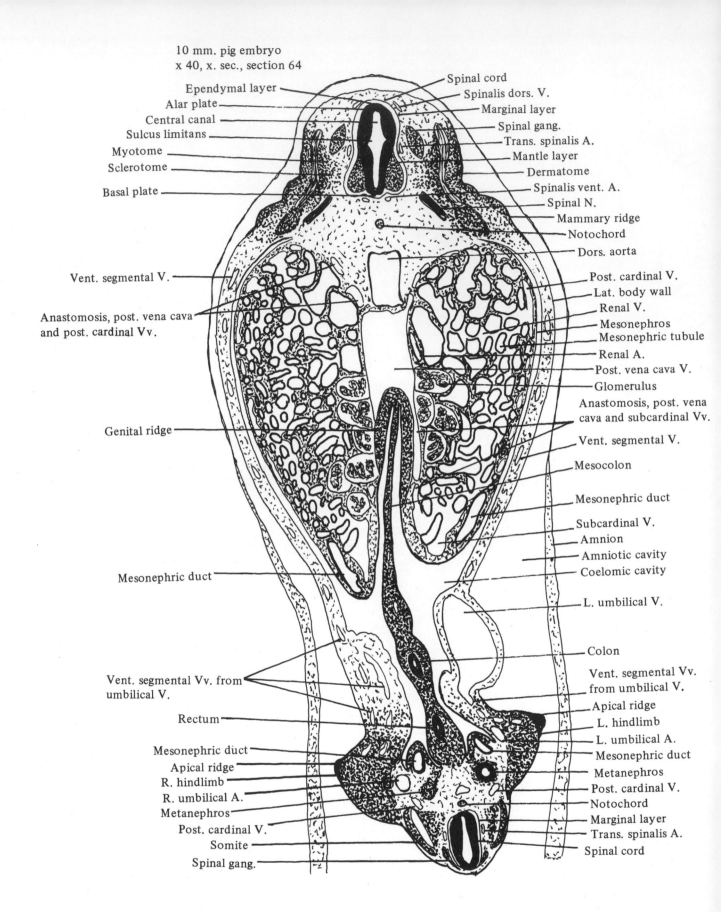

10 mm. pig embryo
x 40, x. sec., section 64

Ependymal layer
Alar plate
Central canal
Sulcus limitans
Myotome
Sclerotome
Basal plate

Spinal cord
Spinalis dors. V.
Marginal layer
Spinal gang.
Trans. spinalis A.
Mantle layer
Dermatome
Spinalis vent. A.
Spinal N.
Mammary ridge
Notochord
Dors. aorta

Vent. segmental V.

Anastomosis, post. vena cava
and post. cardinal Vv.

Post. cardinal V.
Lat. body wall
Renal V.
Mesonephros
Mesonephric tubule
Renal A.
Post. vena cava V.
Glomerulus
Anastomosis, post. vena
cava and subcardinal Vv.

Vent. segmental V.

Genital ridge

Mesocolon

Mesonephric duct

Subcardinal V.
Amnion
Amniotic cavity
Coelomic cavity

Mesonephric duct

L. umbilical V.

Colon

Vent. segmental Vv. from
umbilical V.

Vent. segmental Vv.
from umbilical V.
Apical ridge
L. hindlimb
L. umbilical A.
Mesonephric duct
Metanephros
Post. cardinal V.
Notochord
Marginal layer
Trans. spinalis A.
Spinal cord

Rectum

Mesonephric duct
Apical ridge
R. hindlimb
R. umbilical A.
Metanephros
Post. cardinal V.
Somite
Spinal gang.

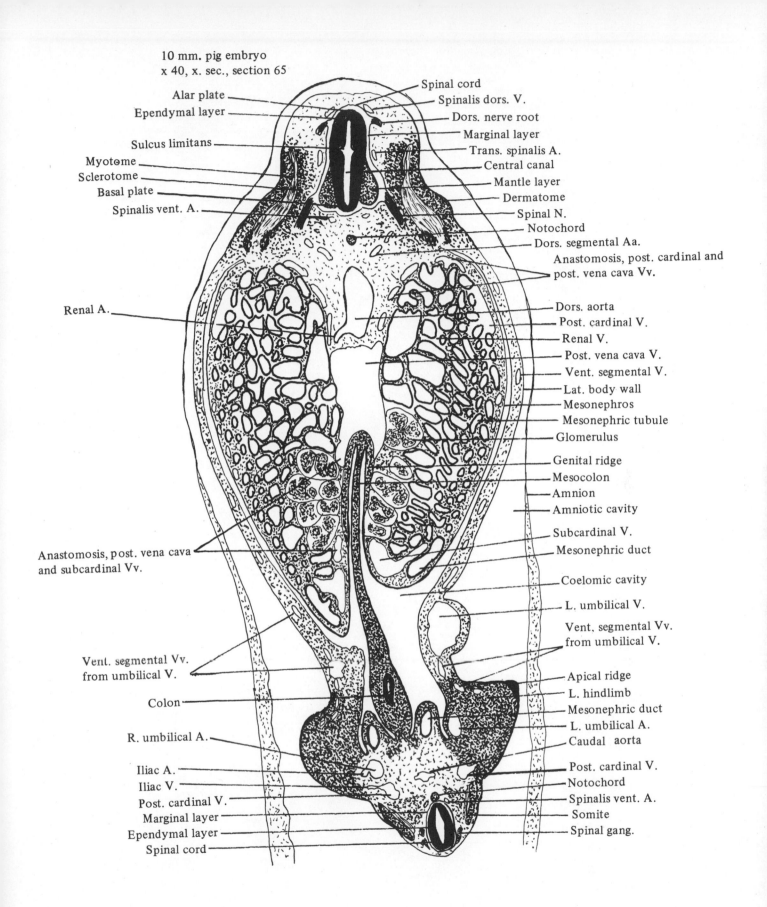

10 mm. pig embryo
x 40, x. sec., section 65

Alar plate
Ependymal layer

Sulcus limitans
Myotome
Sclerotome
Basal plate
Spinalis vent. A.

Spinal cord
Spinalis dors. V.
Dors. nerve root
Marginal layer
Trans. spinalis A.
Central canal
Mantle layer
Dermatome
Spinal N.
Notochord
Dors. segmental Aa.
Anastomosis, post. cardinal and
post. vena cava Vv.

Renal A.

Dors. aorta
Post. cardinal V.
Renal V.
Post. vena cava V.
Vent. segmental V.
Lat. body wall
Mesonephros
Mesonephric tubule
Glomerulus
Genital ridge
Mesocolon
Amnion
Amniotic cavity
Subcardinal V.
Mesonephric duct

Anastomosis, post. vena cava
and subcardinal Vv.

Coelomic cavity
L. umbilical V.
Vent. segmental Vv.
from umbilical V.

Vent. segmental Vv.
from umbilical V.

Colon

R. umbilical A.

Iliac A.
Iliac V.
Post. cardinal V.
Marginal layer
Ependymal layer
Spinal cord

Apical ridge
L. hindlimb
Mesonephric duct
L. umbilical A.
Caudal aorta
Post. cardinal V.
Notochord
Spinalis vent. A.
Somite
Spinal gang.

303

10 mm. pig embryo
x 40, x. sec., section 66

Alar plate

Ependymal layer

Sulcus limitans

Central canal

Dermatome

Myotome

Basal plate

Sclerotome

Notochord

Sympathetic gang.

Post. cardinal V.

Renal Vv.

Post. vena cava V.

Vent. segmental Vv.

Mesonephric duct

R. hindlimb

R. umbilical A.

Iliac A.

Post. cardinal V.

Dors. segmental V.

Sulcus limitans

Central canal

Spinal cord

Spinal cord

Spinalis dors. V.

Marginal layer

Spinal gang.

Trans. spinalis A.

Central canal

Mantle layer

Vent. nerve root

Spinalis vent. A.

Dors. segmental A.

Spinal N.

Dors. aorta

Anastomosis, post. vena cava
and post. cardinal Vv.

Renal A.

Renal Vv.

Lat. body wall

Mesonephros

Mesonephric tubule

Glomerulus

Anastomosis, post. vena
cava and subcardinal Vv.

Subcardinal V.

Mesonephric duct

Amnion

Amniotic cavity

Coelomic cavity

Vent. segmental Vv.
from umbilical V.

Mesocolon

Apical ridge

L. hindlimb

Mesonephric duct

L. umbilical A.

Post. cardinal V.

Caudal aorta

Notochord

Somite

Marginal layer

Spinal gang.

304

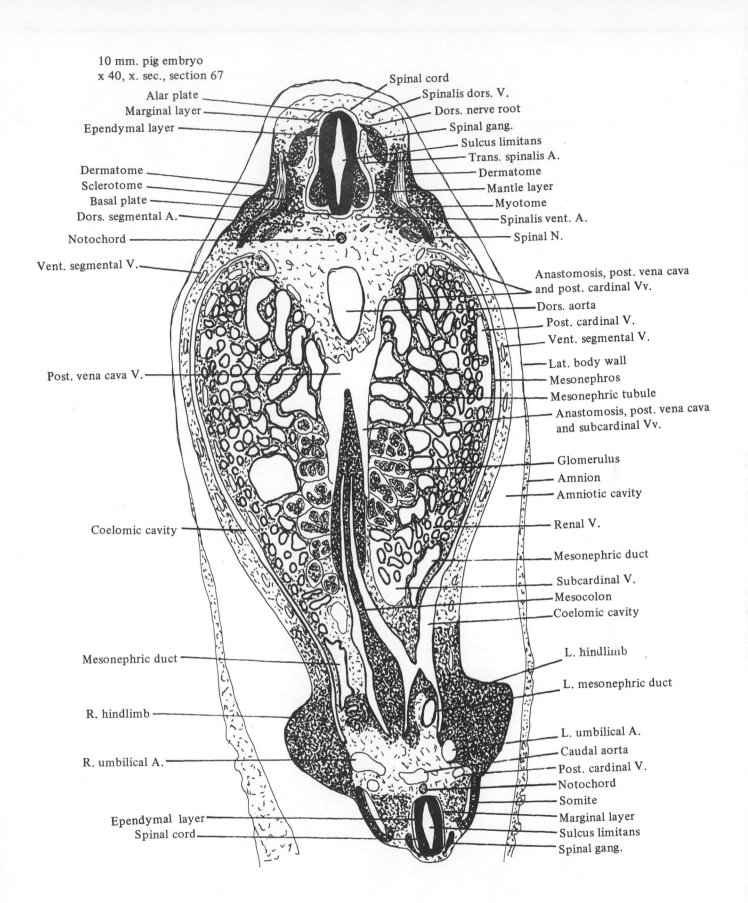

10 mm. pig embryo
x 40, x. sec., section 67

Alar plate
Marginal layer
Ependymal layer

Spinal cord
Spinalis dors. V.
Dors. nerve root
Spinal gang.
Sulcus limitans
Trans. spinalis A.

Dermatome
Sclerotome
Basal plate
Dors. segmental A.
Notochord
Vent. segmental V.

Dermatome
Mantle layer
Myotome
Spinalis vent. A.
Spinal N.

Anastomosis, post. vena cava
and post. cardinal Vv.
Dors. aorta
Post. cardinal V.
Vent. segmental V.

Post. vena cava V.

Lat. body wall
Mesonephros
Mesonephric tubule
Anastomosis, post. vena cava
and subcardinal Vv.

Glomerulus
Amnion
Amniotic cavity

Coelomic cavity

Renal V.

Mesonephric duct
Subcardinal V.
Mesocolon
Coelomic cavity

Mesonephric duct

L. hindlimb

R. hindlimb

L. mesonephric duct

R. umbilical A.

L. umbilical A.
Caudal aorta
Post. cardinal V.
Notochord
Somite
Marginal layer
Sulcus limitans
Spinal gang.

Ependymal layer
Spinal cord

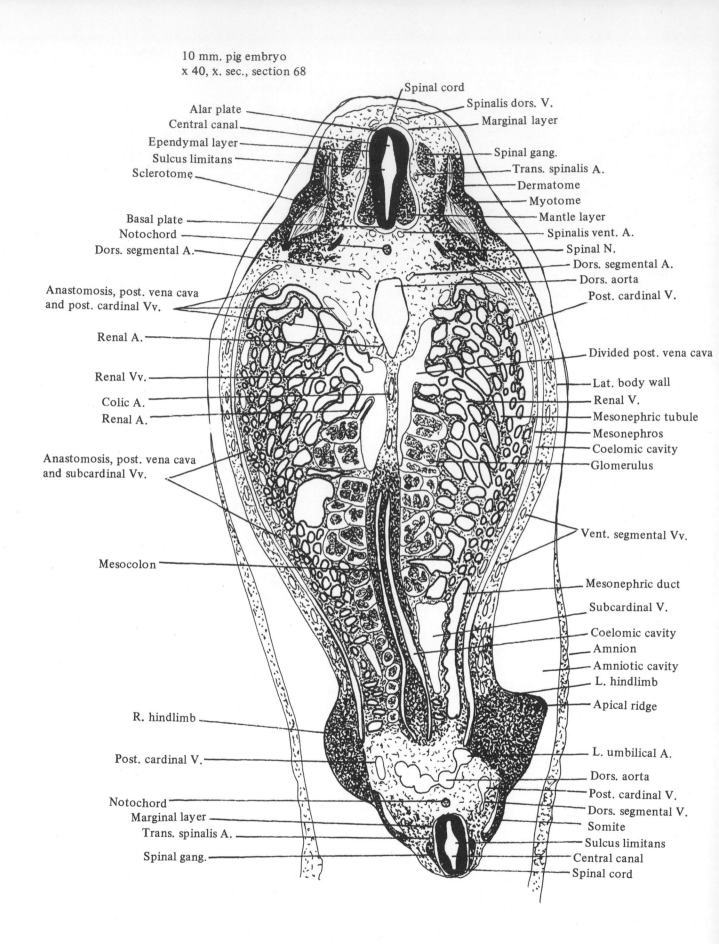

10 mm. pig embryo
x 40, x. sec., section 68

Spinal cord

Alar plate
Central canal
Ependymal layer
Sulcus limitans
Sclerotome

Basal plate
Notochord
Dors. segmental A.

Anastomosis, post. vena cava
and post. cardinal Vv.

Renal A.

Renal Vv.
Colic A.
Renal A.

Anastomosis, post. vena cava
and subcardinal Vv.

Mesocolon

R. hindlimb

Post. cardinal V.

Notochord
Marginal layer
Trans. spinalis A.
Spinal gang.

Spinalis dors. V.
Marginal layer

Spinal gang.
Trans. spinalis A.
Dermatome
Myotome
Mantle layer
Spinalis vent. A.
Spinal N.
Dors. segmental A.
Dors. aorta
Post. cardinal V.

Divided post. vena cava

Lat. body wall
Renal V.
Mesonephric tubule
Mesonephros
Coelomic cavity
Glomerulus

Vent. segmental Vv.

Mesonephric duct
Subcardinal V.
Coelomic cavity
Amnion
Amniotic cavity
L. hindlimb
Apical ridge

L. umbilical A.
Dors. aorta
Post. cardinal V.
Dors. segmental V.
Somite
Sulcus limitans
Central canal
Spinal cord

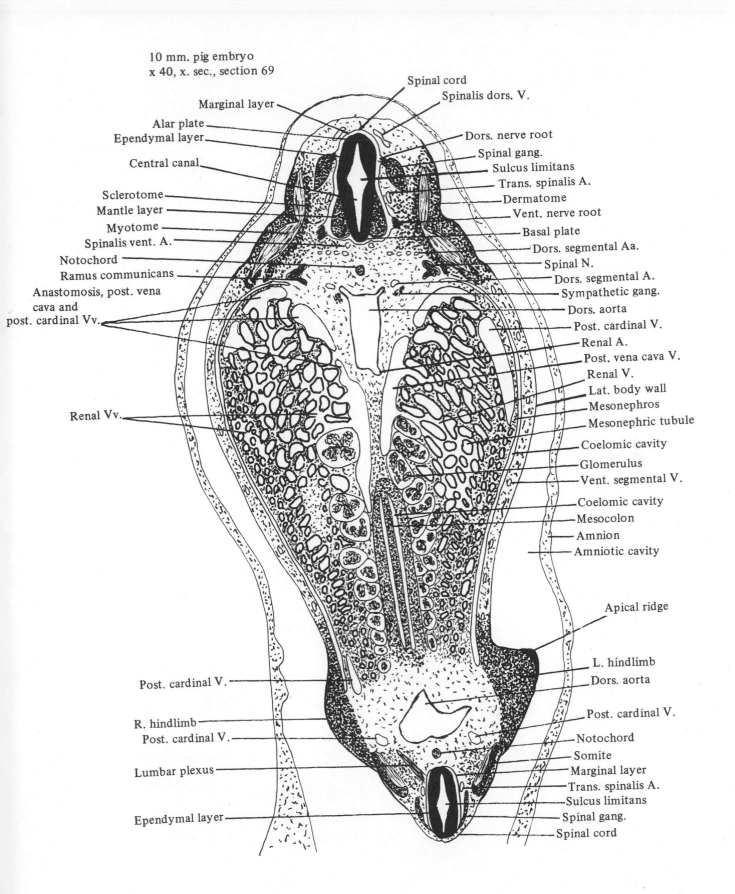

10 mm. pig embryo
x 40, x. sec., section 69

Spinal cord
Spinalis dors. V.
Marginal layer
Alar plate
Ependymal layer
Central canal
Dors. nerve root
Spinal gang.
Sulcus limitans
Trans. spinalis A.
Dermatome
Vent. nerve root
Sclerotome
Mantle layer
Myotome
Spinalis vent. A.
Notochord
Ramus communicans
Basal plate
Dors. segmental Aa.
Spinal N.
Dors. segmental A.
Sympathetic gang.
Anastomosis, post. vena cava and post. cardinal Vv.
Dors. aorta
Post. cardinal V.
Renal A.
Post. vena cava V.
Renal V.
Lat. body wall
Mesonephros
Renal Vv.
Mesonephric tubule
Coelomic cavity
Glomerulus
Vent. segmental V.
Coelomic cavity
Mesocolon
Amnion
Amniotic cavity
Apical ridge
Post. cardinal V.
L. hindlimb
Dors. aorta
R. hindlimb
Post. cardinal V.
Post. cardinal V.
Notochord
Somite
Lumbar plexus
Marginal layer
Trans. spinalis A.
Sulcus limitans
Ependymal layer
Spinal gang.
Spinal cord

10 mm. pig embryo
x 40, x. sec., section 70

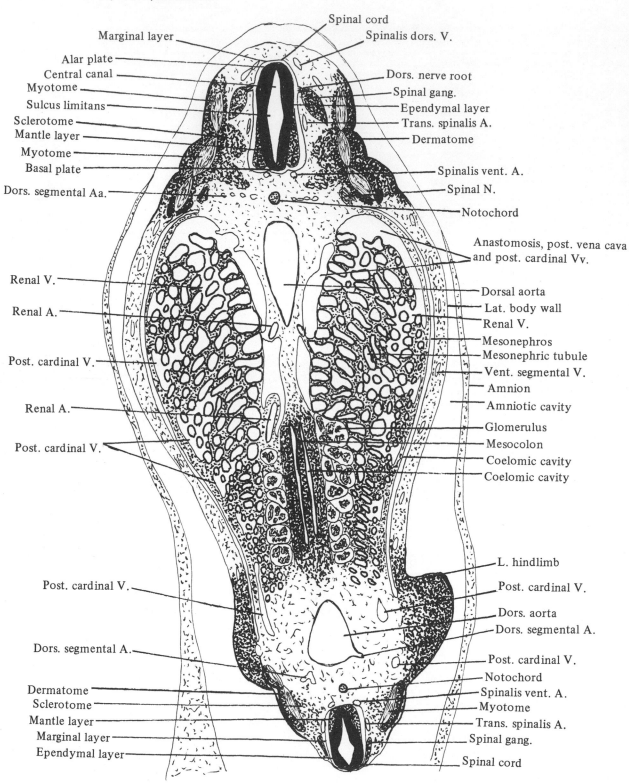

Spinal cord
Spinalis dors. V.
Marginal layer
Alar plate
Central canal
Myotome
Sulcus limitans
Sclerotome
Mantle layer
Myotome
Basal plate
Dors. segmental Aa.

Dors. nerve root
Spinal gang.
Ependymal layer
Trans. spinalis A.
Dermatome
Spinalis vent. A.
Spinal N.
Notochord
Anastomosis, post. vena cava
and post. cardinal Vv.

Renal V.
Renal A.
Post. cardinal V.
Renal A.
Post. cardinal V.

Dorsal aorta
Lat. body wall
Renal V.
Mesonephros
Mesonephric tubule
Vent. segmental V.
Amnion
Amniotic cavity
Glomerulus
Mesocolon
Coelomic cavity
Coelomic cavity

Post. cardinal V.

Dors. segmental A.

Dermatome
Sclerotome
Mantle layer
Marginal layer
Ependymal layer

L. hindlimb
Post. cardinal V.
Dors. aorta
Dors. segmental A.
Post. cardinal V.
Notochord
Spinalis vent. A.
Myotome
Trans. spinalis A.
Spinal gang.
Spinal cord

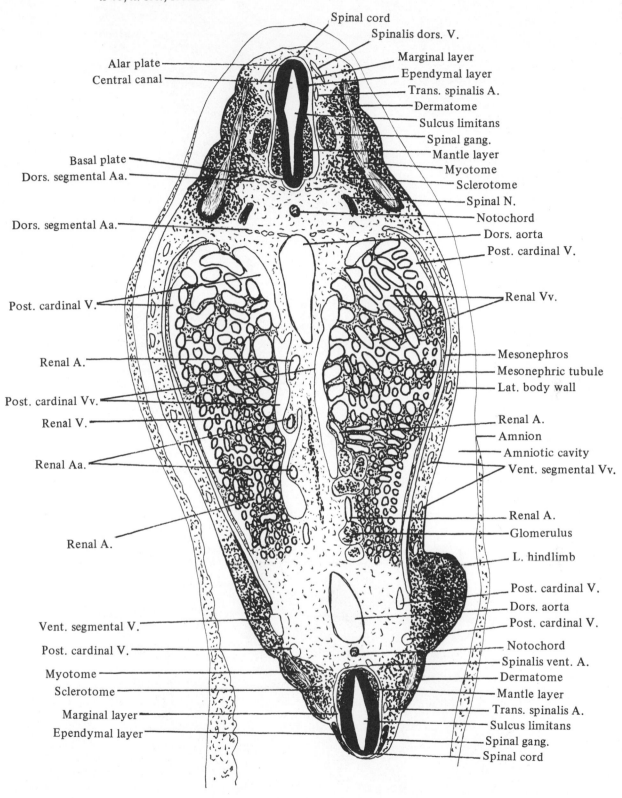

10 mm. pig embryo
x 40, x. sec., section 71

Spinal cord
Spinalis dors. V.
Alar plate
Central canal
Marginal layer
Ependymal layer
Trans. spinalis A.
Dermatome
Sulcus limitans
Spinal gang.
Mantle layer
Basal plate
Myotome
Dors. segmental Aa.
Sclerotome
Spinal N.
Notochord
Dors. segmental Aa.
Dors. aorta
Post. cardinal V.
Post. cardinal V.
Renal Vv.
Renal A.
Mesonephros
Mesonephric tubule
Lat. body wall
Post. cardinal Vv.
Renal V.
Renal A.
Amnion
Amniotic cavity
Renal Aa.
Vent. segmental Vv.
Renal A.
Glomerulus
Renal A.
L. hindlimb
Post. cardinal V.
Dors. aorta
Vent. segmental V.
Post. cardinal V.
Post. cardinal V.
Notochord
Myotome
Spinalis vent. A.
Sclerotome
Dermatome
Marginal layer
Mantle layer
Ependymal layer
Trans. spinalis A.
Sulcus limitans
Spinal gang.
Spinal cord

10 mm. pig embryo
x 40, x. sec., section 72

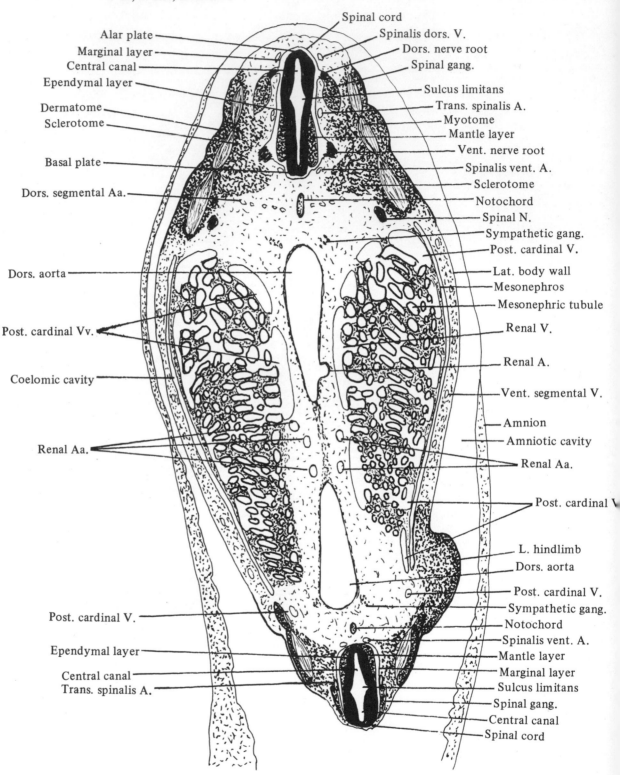

Spinal cord
Alar plate
Marginal layer
Central canal
Ependymal layer
Dermatome
Sclerotome
Basal plate
Dors. segmental Aa.
Dors. aorta
Post. cardinal Vv.
Coelomic cavity
Renal Aa.
Post. cardinal V.
Ependymal layer
Central canal
Trans. spinalis A.

Spinalis dors. V.
Dors. nerve root
Spinal gang.
Sulcus limitans
Trans. spinalis A.
Myotome
Mantle layer
Vent. nerve root
Spinalis vent. A.
Sclerotome
Notochord
Spinal N.
Sympathetic gang.
Post. cardinal V.
Lat. body wall
Mesonephros
Mesonephric tubule
Renal V.
Renal A.
Vent. segmental V.
Amnion
Amniotic cavity
Renal Aa.
Post. cardinal V
L. hindlimb
Dors. aorta
Post. cardinal V.
Sympathetic gang.
Notochord
Spinalis vent. A.
Mantle layer
Marginal layer
Sulcus limitans
Spinal gang.
Central canal
Spinal cord

310

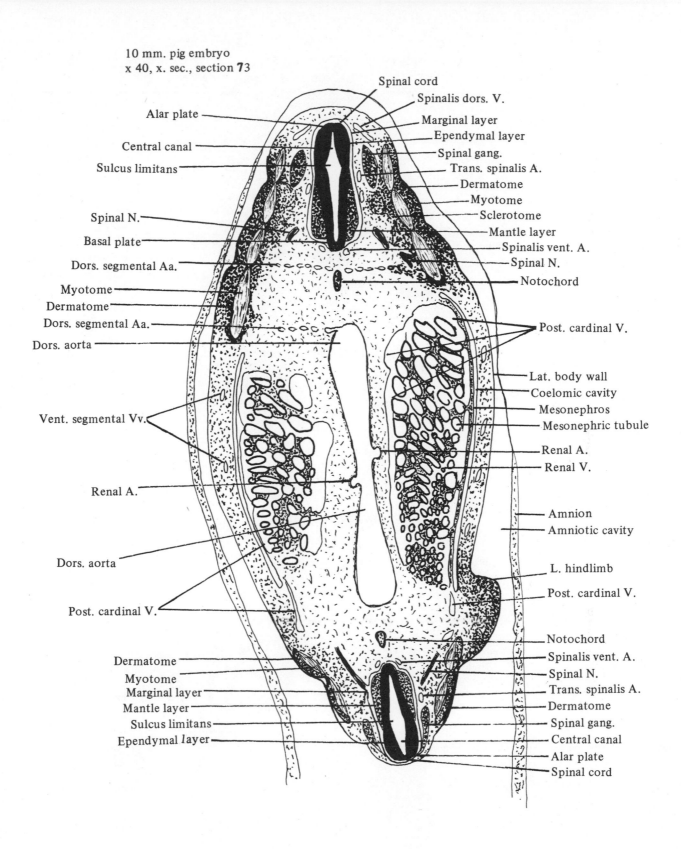

10 mm. pig embryo
x 40, x. sec., section 73

Spinal cord
Spinalis dors. V.
Alar plate
Marginal layer
Ependymal layer
Central canal
Spinal gang.
Sulcus limitans
Trans. spinalis A.
Dermatome
Myotome
Sclerotome
Spinal N.
Mantle layer
Basal plate
Spinalis vent. A.
Dors. segmental Aa.
Spinal N.
Myotome
Notochord
Dermatome
Dors. segmental Aa.
Post. cardinal V.
Dors. aorta
Lat. body wall
Coelomic cavity
Mesonephros
Mesonephric tubule
Vent. segmental Vv.
Renal A.
Renal V.
Renal A.
Amnion
Amniotic cavity
Dors. aorta
L. hindlimb
Post. cardinal V.
Post. cardinal V.
Notochord
Dermatome
Spinalis vent. A.
Myotome
Spinal N.
Marginal layer
Trans. spinalis A.
Mantle layer
Dermatome
Sulcus limitans
Spinal gang.
Ependymal layer
Central canal
Alar plate
Spinal cord

10 mm. pig embryo
x 40, x. sec., section 74

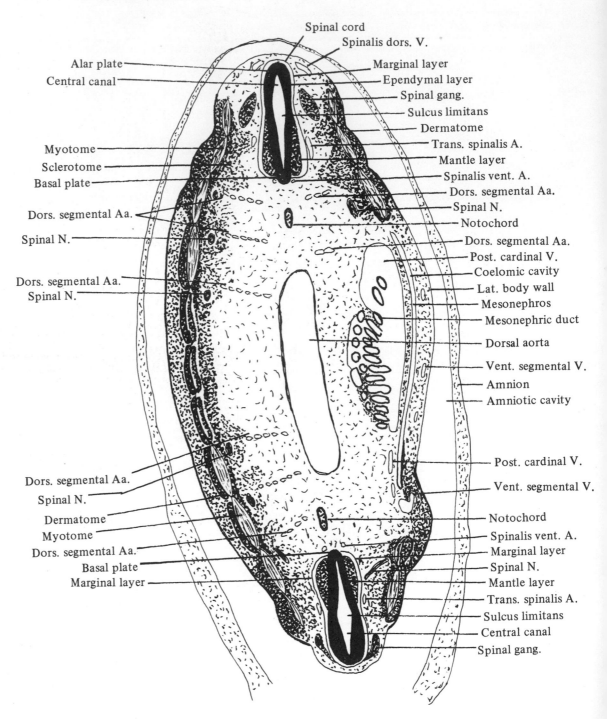

Spinal cord
Spinalis dors. V.
Alar plate
Central canal
Marginal layer
Ependymal layer
Spinal gang.
Sulcus limitans
Dermatome
Myotome
Trans. spinalis A.
Sclerotome
Mantle layer
Basal plate
Spinalis vent. A.
Dors. segmental Aa.
Dors. segmental Aa.
Spinal N.
Spinal N.
Notochord
Dors. segmental Aa.
Post. cardinal V.
Coelomic cavity
Dors. segmental Aa.
Lat. body wall
Spinal N.
Mesonephros
Mesonephric duct
Dorsal aorta
Vent. segmental V.
Amnion
Amniotic cavity

Post. cardinal V.
Dors. segmental Aa.
Vent. segmental V.
Spinal N.
Dermatome
Notochord
Myotome
Spinalis vent. A.
Dors. segmental Aa.
Marginal layer
Basal plate
Spinal N.
Marginal layer
Mantle layer
Trans. spinalis A.
Sulcus limitans
Central canal
Spinal gang.

10 mm. pig embryo
x 40, x. sec., section 75

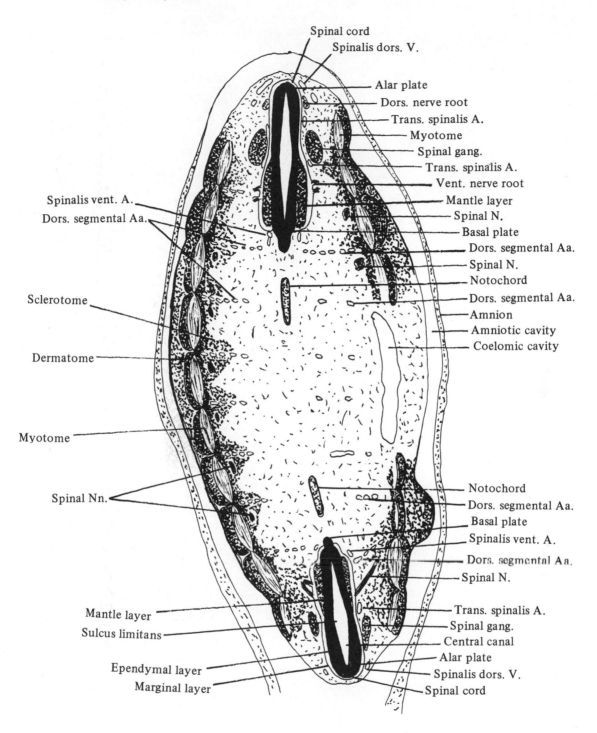

Spinal cord
Spinalis dors. V.
Alar plate
Dors. nerve root
Trans. spinalis A.
Myotome
Spinal gang.
Trans. spinalis A.
Vent. nerve root
Mantle layer
Spinal N.
Basal plate
Dors. segmental Aa.
Spinal N.
Notochord
Dors. segmental Aa.
Amnion
Amniotic cavity
Coelomic cavity

Spinalis vent. A.
Dors. segmental Aa.

Sclerotome

Dermatome

Myotome

Spinal Nn.

Notochord
Dors. segmental Aa.
Basal plate
Spinalis vent. A.
Dors. segmental Aa.
Spinal N.
Trans. spinalis A.
Spinal gang.
Central canal
Alar plate
Spinalis dors. V.
Spinal cord

Mantle layer
Sulcus limitans
Ependymal layer
Marginal layer

313

10 mm. pig embryo
x 40, x. sec., section 76

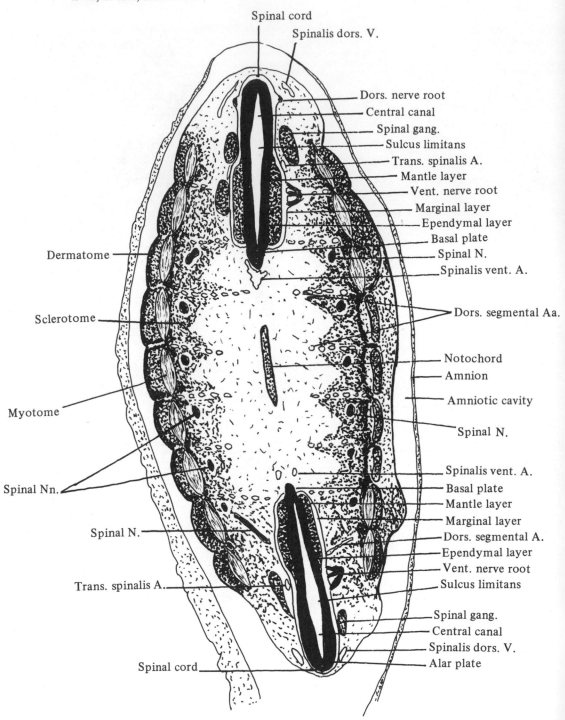

Spinal cord

Spinalis dors. V.

Dors. nerve root
Central canal
Spinal gang.
Sulcus limitans
Trans. spinalis A.
Mantle layer
Vent. nerve root
Marginal layer
Ependymal layer
Basal plate
Spinal N.
Spinalis vent. A.

Dermatome

Dors. segmental Aa.

Sclerotome

Notochord
Amnion

Amniotic cavity

Myotome

Spinal N.

Spinalis vent. A.
Basal plate
Mantle layer
Marginal layer
Dors. segmental A.
Ependymal layer
Vent. nerve root
Sulcus limitans

Spinal Nn.

Spinal N.

Trans. spinalis A.

Spinal gang.
Central canal
Spinalis dors. V.
Alar plate

Spinal cord

314

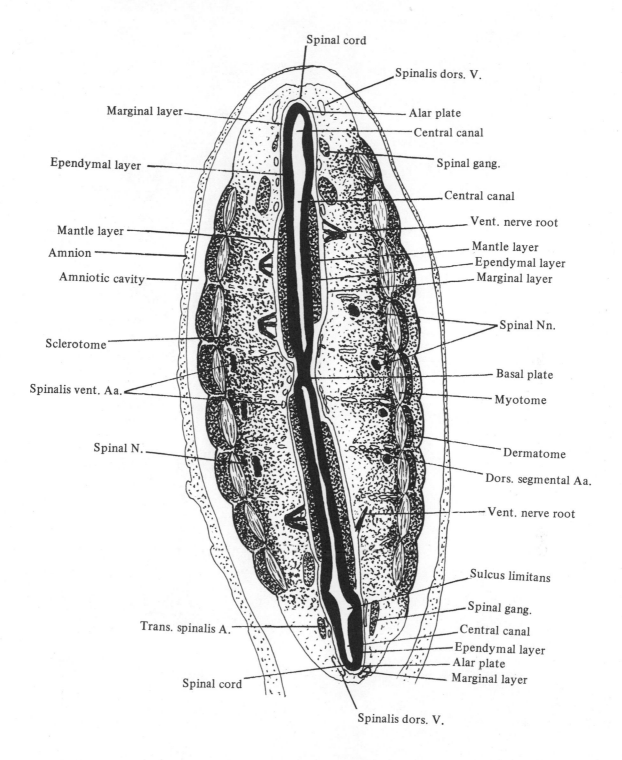

10 mm. pig embryo
x 40, x. sec., section 77

Spinal cord

Spinalis dors. V.

Marginal layer

Alar plate

Central canal

Ependymal layer

Spinal gang.

Central canal

Mantle layer

Vent. nerve root

Amnion

Mantle layer

Amniotic cavity

Ependymal layer

Marginal layer

Spinal Nn.

Sclerotome

Spinalis vent. Aa.

Basal plate

Myotome

Spinal N.

Dermatome

Dors. segmental Aa.

Vent. nerve root

Sulcus limitans

Spinal gang.

Trans. spinalis A.

Central canal

Ependymal layer

Alar plate

Spinal cord

Marginal layer

Spinalis dors. V.

10 mm. pig embryo
x 40, x. sec., section 78

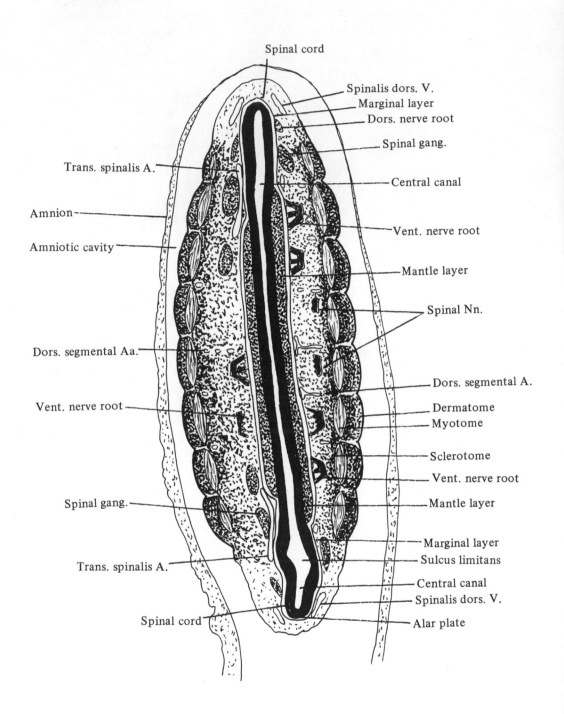

Spinal cord

Spinalis dors. V.

Marginal layer

Dors. nerve root

Spinal gang.

Central canal

Trans. spinalis A.

Amnion

Amniotic cavity

Vent. nerve root

Mantle layer

Spinal Nn.

Dors. segmental Aa.

Dors. segmental A.

Dermatome

Myotome

Vent. nerve root

Sclerotome

Vent. nerve root

Spinal gang.

Mantle layer

Marginal layer

Sulcus limitans

Trans. spinalis A.

Central canal

Spinalis dors. V.

Spinal cord

Alar plate

10 mm. pig embryo
x 40, x. sec., section 79

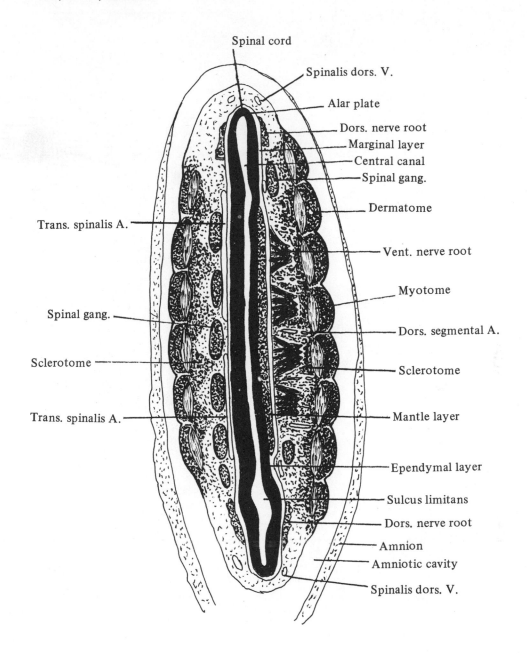

Spinal cord

Spinalis dors. V.

Alar plate

Dors. nerve root

Marginal layer

Central canal

Spinal gang.

Dermatome

Trans. spinalis A.

Vent. nerve root

Myotome

Spinal gang.

Dors. segmental A.

Sclerotome

Sclerotome

Trans. spinalis A.

Mantle layer

Ependymal layer

Sulcus limitans

Dors. nerve root

Amnion

Amniotic cavity

Spinalis dors. V.

10 mm. pig embryo
x 40, x. sec., section 80

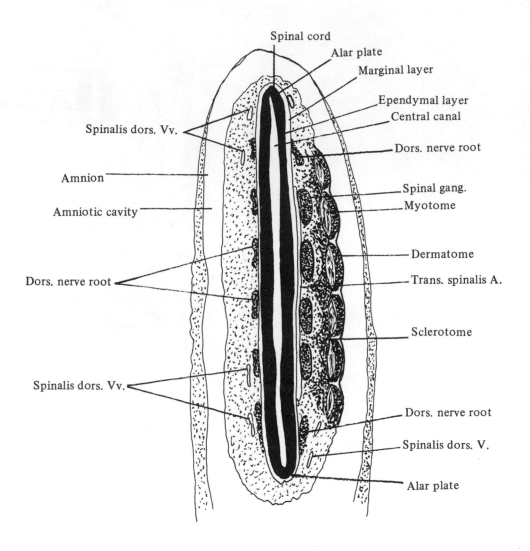

Spinal cord

Alar plate

Marginal layer

Ependymal layer

Central canal

Dors. nerve root

Spinalis dors. Vv.

Amnion

Amniotic cavity

Spinal gang.

Myotome

Dermatome

Trans. spinalis A.

Dors. nerve root

Sclerotome

Spinalis dors. Vv.

Dors. nerve root

Spinalis dors. V.

Alar plate

10 mm. pig embryo
x 40, x. sec., section 81

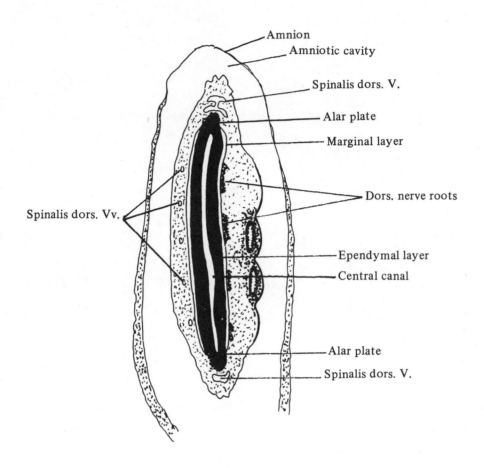

Amnion
Amniotic cavity
Spinalis dors. V.
Alar plate
Marginal layer
Dors. nerve roots
Spinalis dors. Vv.
Ependymal layer
Central canal
Alar plate
Spinalis dors. V.

Mammalian Development III

More-Developed 10 mm. Pig Embryo

10 mm. pig embryo
x 60, x. sec., section 1

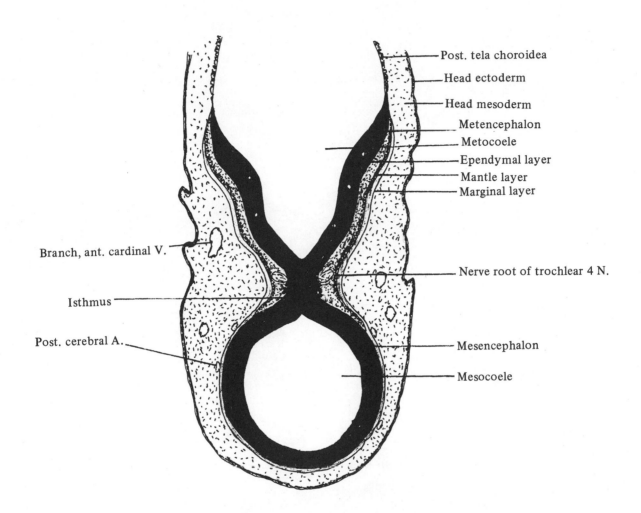

Post. tela choroidea

Head ectoderm

Head mesoderm

Metencephalon

Metocoele

Ependymal layer

Mantle layer

Marginal layer

Branch, ant. cardinal V.

Nerve root of trochlear 4 N.

Isthmus

Post. cerebral A.

Mesencephalon

Mesocoele

10 mm. pig embryo
x 60, x. sec., section 2

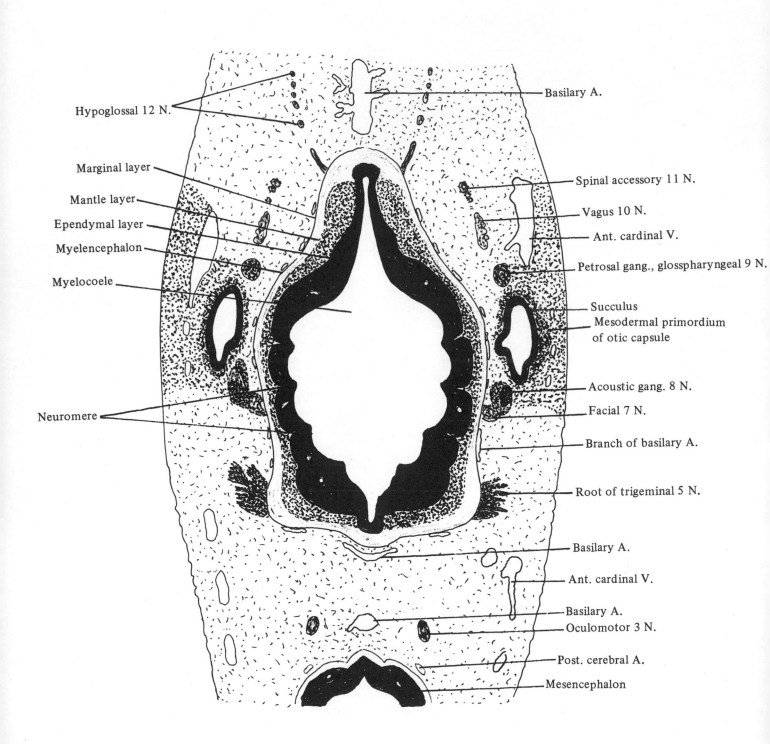

Hypoglossal 12 N.

Marginal layer

Mantle layer

Ependymal layer

Myelencephalon

Myelocoele

Neuromere

Basilary A.

Spinal accessory 11 N.

Vagus 10 N.

Ant. cardinal V.

Petrosal gang., glosspharyngeal 9 N.

Succulus
Mesodermal primordium
of otic capsule

Acoustic gang. 8 N.

Facial 7 N.

Branch of basilary A.

Root of trigeminal 5 N.

Basilary A.

Ant. cardinal V.

Basilary A.
Oculomotor 3 N.

Post. cerebral A.

Mesencephalon

10 mm. pig embryo
x 60, x. sec., section 3

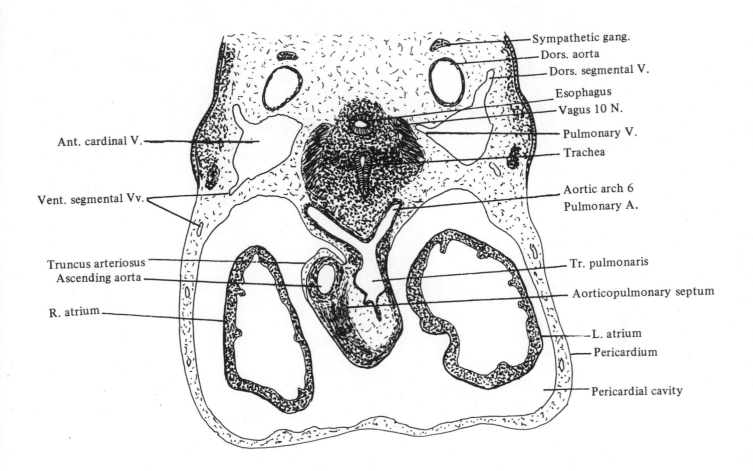

Sympathetic gang.
Dors. aorta
Dors. segmental V.
Esophagus
Vagus 10 N.
Pulmonary V.
Trachea
Aortic arch 6
Pulmonary A.
Tr. pulmonaris
Aorticopulmonary septum
L. atrium
Pericardium
Pericardial cavity

Ant. cardinal V.

Vent. segmental Vv.

Truncus arteriosus
Ascending aorta
R. atrium

10 mm. pig embryo
x 60, x. sec., section 4

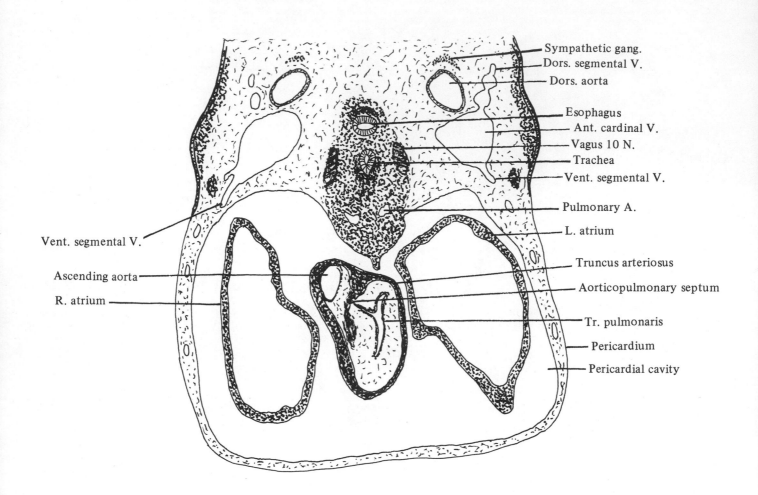

Sympathetic gang.
Dors. segmental V.
Dors. aorta

Esophagus
Ant. cardinal V.
Vagus 10 N.
Trachea
Vent. segmental V.

Pulmonary A.

L. atrium

Vent. segmental V.

Truncus arteriosus

Ascending aorta

Aorticopulmonary septum

R. atrium

Tr. pulmonaris
Pericardium
Pericardial cavity

10 mm. pig embryo
x 60, x. sec., section 5

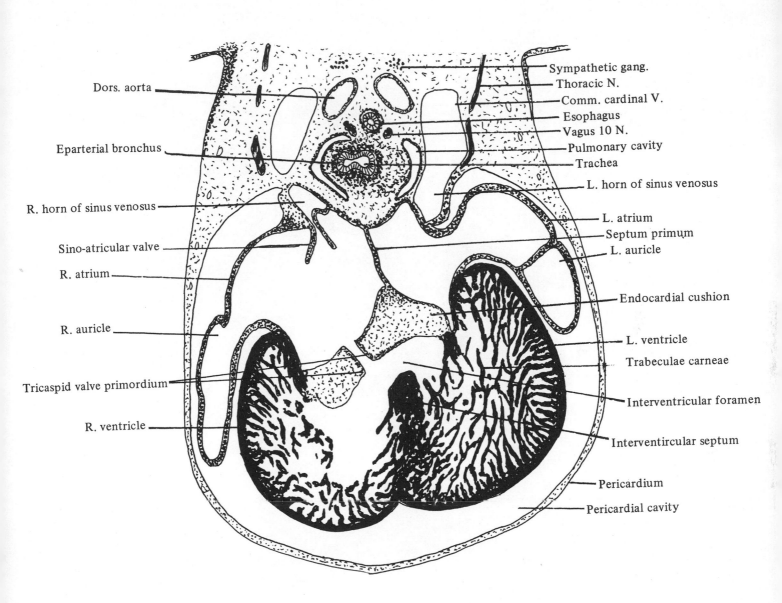

Dors. aorta

Eparterial bronchus

R. horn of sinus venosus

Sino-atricular valve

R. atrium

R. auricle

Tricaspid valve primordium

R. ventricle

Sympathetic gang.
Thoracic N.
Comm. cardinal V.
Esophagus
Vagus 10 N.
Pulmonary cavity
Trachea
L. horn of sinus venosus

L. atrium
Septum primum
L. auricle

Endocardial cushion

L. ventricle
Trabeculae carneae

Interventricular foramen

Interventircular septum

Pericardium
Pericardial cavity

10 mm. pig embryo
x 60, x. sec., section 6

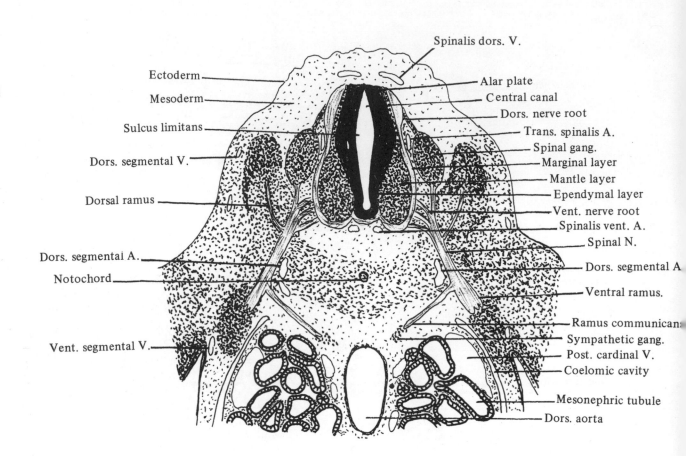

Spinalis dors. V.

Ectoderm

Alar plate

Mesoderm

Central canal

Sulcus limitans

Dors. nerve root

Trans. spinalis A.

Dors. segmental V.

Spinal gang.

Marginal layer

Mantle layer

Dorsal ramus

Ependymal layer

Vent. nerve root

Spinalis vent. A.

Spinal N.

Dors. segmental A.

Dors. segmental A

Notochord

Ventral ramus.

Ramus communican

Sympathetic gang.

Vent. segmental V.

Post. cardinal V.

Coelomic cavity

Mesonephric tubule

Dors. aorta

Appendix

Terminology Used in the Text

ATLAS OF DEVELOPMENTAL
ANATOMY

PART I

SPERMATOGENESIS
Anaphase I
Apex of testicular lobe
Cellular membrane of the testicular lobe
Cellular septum
Cyst
Diakinesis
Diplotene
Ductuli efferentes
Early spermatid
First maturation division
Interlobular space
Leptotene
Metaphase I
Metaphase I. polar view
Mitosis of spermatogonia
Nucleolus
Nurse cell
Pachytene
Premetaphase
Primary spermatocyte
Septum
Spermatozoon
Second maturation division
Secondary spermatocyte
Spermatid
Spermiogenesis
Spermatogonia
Synaptene (Zygotene)
Tail formation of the spermatid
Telophase I
Testicular lobe
Tunica albuginea
Vasa deferens
Zygotene (Synaptene)

OOGENESIS
Activated sperm
Anaphase
Anaphase of the first maturation division
Aster formation
Bilateral cleavage
Early anaphase
Early metaphase
Early telophase
Female genital tract
Female nucleus
Fertilization membrane
First maturation division
First polar body
Four cell stage
Germinal zone
Male pronucleus
Metaphase
Ovary
Ovum
Oogonia
Oogonium
Perivitelline space
Primary oocyte
Pronuclei with chromosome formation
Pronculeus
Second maturation division
Secondary oocyte
Second polar body
Shell
Sperm
Sperm attachment
Sperm nuclei
Sperm nucleus
Sperm penetration
Telophase
Telophase of the first maturation division
Telophase of the second maturation
    division
Terminal zone
Tetrad formation
Two cell stage
Uterus

Vagina
Vulva
Wall of the uterus

EARLY DEVELOPMENT OF
THE STARFISH
Animal pole
Archenteric canal
Archenteric vesicle
Archenteron
Beginning of gastrulation
Blastocoele
Blastoderm
Blastomere
Blastopore
Early blastula
Early gastrula
Ectoderm
Eight-cell stage
Endoderm
Fertilization membrane
First cleavage
Four-cell stage
Fourth cleavage
Late blastula
Late gastrula
Mesenchyme cell
Mesenchyme cell formation
Mesothelium
Mitosis
Morula
Nucleolus
Nucleus
Oogonium
Polar body
Second cleavage
Sixteen-cell stage
Third cleavage
Two-cell stage
Vegetal pole
Vitelline membrane

327

DEVELOPMENT OF AMPHIOXUS

Alimentary canal
Anterior
Anus
Archenteron
Atriopore
Atrium
Blastopore
Brain
Buccal cavity
Caudal fin
Cirri
Cranial nerves
Dorsal
Dors. fin ray
Dorsocaudal fin
Early blastula
Early larva
Ectoderm
Eight-cell stage
Endoderm
Endostyle
Epibranchial gooove
Eye spot
First cleavage
Fertilization membrane
Four-cell stage
Fourth cleavage
Gastrula
Gill
Gland
Iliocolon ring
Immature adult
Late embryo
Late larva
Lateral diverticulum
Lat. plate mesoderm
Liver
Liver primordium
Metapleural fold
Morula
Myomere
Myoseptum
Neural canal
Neural tube
Neuro pore
Neurenteric canal
Notochord
Pharynx
Pigment spot
Posterior
Rostrum
Second cleavage
Sixteen-cell stage
Somites
Spinal cord
Third cleavage
Transverse muscle
Two cell stage

Uncleaved egg
Unfertilized egg
Velar tenticle
Velum
Ventral
Vent. fin ray
Ventrocaudal fin
Vitelline membrane
Wheel organ

DEVELOPMENT OF WHITEFISH

Acoustic N. 8
Alar plate
Ant. cardinal Vv.
Basal plate
Blood island
Central canal
Coelomic cavity
Dermatome
Dielencephalon
Dors. aorta
Dors. fin fold
Dors. mesentery
Early cleavage
Early metaphase
Early prophase
Ectoderm
Endocardium
Endoderm
Epaxial myomere
Ependymal layer
Epithelial layer
Epithelial layer of ectoderm
Erythrocytes
Extraembryonic tissue
Eye cup
Facial N. 7
First cleavage
First somite
Gastrula
Gut
Gut cavity
Head ectoderm
Head fold
Head mesenchyme
Heart
Hypaxial myomere
Isthmus
Keel
Late prophase
Lateral line
Lat. plate of mesoderm
Lens
Mesencephalon
Mesenchyme
Metaphase II
Mesocoele
Mesoderm
Metencephalon
Mitotic figures

Myelencephalon
Myelocoele
Myocardium
Myotome
Nasal placode
Neural layer
Neural layer of ectoderm
Neural tube
Neuroectoderm
Neuroectoderm-keel
Neuromere of myelencephalon
Neurulation
Notochord
Notochordal sheath
Oogenesis
Optic lobe
Optic vesicle
Otic capsule
Otic vesicle
Pectoral fin
Periblast
Periderm
Pharyngeal pouches
Post. end of subcephalic pocket
Post. eye chamber
Pronephric bud
Pronephric duct
Pronephric tubule
Pronuclei
Prosencephalon
Rhombencephalon
Roof of the myelencephalon
Sclerotome
Second cleavage
Sensory layer
Separation slit
Spinal cord
Subcephalic pocket
Tail fin fold
Tail rudiment
Tapetum nigrum
Telencephalon
Telophase II
Tenth somite
Trigeminal N. 5
Urinary bladder
Vent. mesentery
Vitelline V.
Yolk
Yolk sac

FROG DEVELOPMENT I
FROG OVARY

Capillaris
Endothelium
Erythrocytes
Follicular cells
Nucleolus
Nucleus
Pigment cells

Pigmented animal hemisphere
Theca externa
Theca interna
Vegetal hemisphere with yolk platelets
Vitelline membrane
Young oocyte

FROG TESTIS
Anaphase
Basement membrane
Blood vessel
Collecting tubule
Collecting tubules, Spring
Collecting tubules, Winter
Diakinesis
Early spermatid
Erythrocytes
Fat tissue
Intersticial space
Intersticial tissue
Late spermatid
Leptotene
Lumen of seminiferous tubule
Mesentery of the testis
Mesorchium
Metaphase
Nucleus of Sertoli cell
Pachytene
Second maturation division
Secondary spermatocytes
Seminoferous tubules
Spermatogonium
Spermatozoa
Synaptene
Tunica albuginea
Vasa deferens

MATURED FROG EGG
Follicular cells
Nucleolus
Nucleus
Pigmented animal pole
Theca externa
Theca interna
Vegetal pole
Vitelline membrane

EARLY CLEAVAGE, EIGHT – CELL STAGE
Animal pole
Fertilization membrane
Third cleavage furrow
Vegetal pole

LATE CLEAVAGE, EARLY BLASTULA
Animal hemisphere
Blastocoele
Blastomere
Fertilization membrane

Pigment
Vegetal hemisphere
Yolk cell

LATE BLASTULA
Animal pole
Blastocoele
Blastoderm
Blastomere
Fertilization membrane
Marginal zone
Vegetal pole
Yolk cell

EARLY AND LATE GASTRULA
Animal pole
Anterior
Archenteron
Blastocoele
Blastopore
Completion bridge
Dorsal
Dors. lip of blastopore
Endoderm
Epithelial layer
Fertilization membrane
Gastrular slit
Involuting cells
Neural layer
Notochord
Peristomial mesoderm
Posterior
Vegetal pole
Ventral
Ventral lip of the blastopore
Yolk cell
Yolk plug

NEURAL PLATE
Chordomesodermal mantle
Ectoderm
Endoderm
Forgut
Mesoderm
Midgut
Neural crest
Neural plate, ant. section
Neural (medullary) plate
Neural plate, post. section
Notochord
Stomodeal invagination
Yolk cell

FROG DEVELOPMENT II
FROG; NEURAL FOLD
Blastopore
Brain area

Dors. diverticulum
Endoderm
Epimere
Epithelial layer
Fertilization membrane
Foregut
Lat. plate mesoderm
Liver diverticulum
Hindgut
Hypomere
Mesenchyme
Mesomere
Midgut
Neural fold
Neural groove
Neural layer
Neuroectoderm
Notochord
Oral diverticulum
Two layered ectoderm
Vent. diverticulum
Yolk cell

3 MM. FROG EMBRYO
Ant. horn of the pharynx
Auditory placode
Central canal
Coelom
Dors. diverticulum
Ectoderm
Endoderm
Eye vesicle
Foregut
Gill rudiment
Head mesenchyme
Heart rudiment
Heart mesoderm
Hindgut
Hypophysis
Infundibulum
Lat. plate mesoderm
Liver
Liver diverticulum
Mesencephalon
Mesocoele
Mesoderm
Midgut
Neural crest
Neural tube
Notochord
Oral sucker
Otic vesicle
Pharynx
Proctodeum
Pronephros
Prosencephalon
Prosocoele
Rhombencephalon
Rhombocoele

Somite
Spinal cord
Stalk of somite
Subnotochordal rod
Tuberculum post.
Two layered ectoderm
Vent. diverticulum
Yolk cell

FROG DEVELOPMENT III
4 MM. FROG EMBRYO
Alar plate
Anus
Basal plate
Central canal
Cloaca, hindgut
Coelom
Dors. fin
Ectoderm
Endocardium
Endoderm
Epiphysis
First somite
Foregut
Gut endoderm
Head mesenchyme
Hindgut
Hypophysis
Infundibulum
Lat. plate mesoderm
Liver diverticulum
Mesenchyme
Mesencephalon
Mesocoele
Mesoderm
Midgut
Myocardium
Nasal placode
Neural crest
Neurocoele
Notochord
Pharynx
Pharyngeal pouch
Pericardial cavity
Pericardium
Pronephric tissue
Prosencephalon
Prosocoele
Omphalomesenteric V.
Oral diverticulum
Oral sucker
Otic vesicle
Rhombencephalon
Rhombocoele
Sensory layer of the eye
Splanchnopleura
Spinal cord
Splanchnic mesoderm
Somite formation

Somatopleura
Stomodeum
Subnotochordal rod
Tail fin
Tail somite
Tapetum nigrum
Trigeminal N. 5
Vent. aorta
Vent. fin
Visceral arch
Yolk cell

7 MM. FROG EMBRYO
Acoustic N. 8
Afferent branchial A.
Ant. cardinal V.
Anus
Atrium
Basilary A.
Bronchus
Cartilage
Cauda
Caudal A.
Caudal V.
Central canal
Choroid fissure
Cloaca
Coelom
Comm. cardinal Vv.
Diencephalon
Diocoele
Dors. aorta
Dors. caudal fin
Dors. fin
Dors. mesentery
Ecotderm
Efferent branchial A.
Endoderm
Endodermal plug
Endolymphatic duct
Ependymal layer
Epiphysis
Esophagus
Esophagus with cellular plug
Ext. gills
External naris
First somite
Facial N. 7
Gall bladder
Genital ridge
Gill stalk
Glossopharyngeal N. 9
Hepatic portal V.
Hindgut
Hyoid cartilage
Hypophysis
Infundibulum
Intestine
Int. carotid A.

Int. glomerulus
Int. naris
Lat. appendix
Larynx
Lens
Liver
Liver cord
Lung bud
Mandible
Mandibular N.
Mantle layer
Marginal layer
Maxillary N.
Mesencephalon
Mesenchyme
Mesenteric V.
Mesocoele
Mesonephric duct
Mesonephric tubules
Metencephalon
Metocoele
Mouth
Myelencephalon
Myelocoele
Neural crest
Notochord
Pancreas
Peribronchial cavity
Pericardial cavity
Pericardium
Pharynx
Post. cardinal V.
Post. end of lung bud
Post. tela choroidea
Recessus opticus
Rectum
Olfactory epithelium
Omphalomesenteric A.
Omphalomesenteric V.
Optic N. 2
Oral cavity
Oral papilla
Oral sucker
Oro-pharyngeal membrane
Otic vesicle
Sensory layer
Sinus venosus
Somite
Spinal cord
Spinal gang.
Subcardinal V.
Subnotochordal rod
Tapetum nigrum
Telencephalon
Telocoele
Thyreoid gl.
Trachea
Transverse septum
Trigeminal N. 5
Truncus arteriosus

Lens placode
Mesencephalon
Mesenchyme
Mesocardium
Mesocoele
Mesoderm
Neural crest
Neural fold
Notochord
Omphalomesenteric V.
Optic vesicle
Oropharyngeal membrane
Otic placode
Pharynx
Pericardium
Pericardial coelomic cavity
Post. end of subcephalic pocket
Post. neuro pore
Primitive streak
Proamnion
Pronephros
Prosencephalon
Prosocoele
Rhombencephalon
Rhombocoele
Sinus rhomboidalis
Sinus terminalis
Sinus venosus
Somotapleura
Somite
Spinal cord
Splanchnopleura
Stalk of somite
Stomodeum
Subcephalic pocket
Thyreoid rudiment
Truncus arteriosus
Tuberculum post.
Vent. aorta
Ventricle
Yolk sac

DEVELOPMENT OF
THE CHICK EMBRYO V

48-HOUR CHICK EMBRYO
Acoustic gang. 8 N.
Amnion
Amniotic cavity
Amniotic fold
Ant. cardinal V.
Ant. end of otic vesicle
Ant. end of telencephalon
Ant. intestinal portal
Ant. vitelline V.
Aortic arch 1
Aortic arch 2
Aortic arch 3
Atrium

Branchial arch 1 (mandibular arch)
Branchail groove 1
Caudal aorta
Caudal fold
Central canal
Cephalic flexure
Chorion
Chorio-amniotic raphe
Choroid fissure
Closing plate
Comm. cardinal V.
Diencephalon
Diocoele
Dors. aorta
Dors. liver diverticulum
Duct of Cuvier
Ductus venosus
Duodenum
Ectoderm
Embryonic coelom
Endocardium
End of spinal cord
Epimyocardium
Ext. embryonic coelom
Foregut
Semilunar gang. N. 5
Geniculate gang. N. 7
Hindgut
Head ectoderm
Head mesoderm
Infundibulum
Int. carotid A.
Intersegmental V.
Isthmus
Lat. body fold
Lat. groove of stomodeum
L. omphalomesenteric A.
L. omphalomesenteric V.
L. vitelline V.
Lens
Lens placode
Maxillary process
Mesencephalon
Mesocardium
Mesocoele
Mesoderm
Mesonephric duct
Mesonephric tubule
Metencephalon
Metocoele
Myelencephalon
Myelocoele
Neural fold
Notochod
Omphalomesenteric A.
Omphalomesenteric V.
Optic cup
Optic stalk
Oropharyngeal membrane
Otic vesicle

Paired dors. aorta
Pharynx
Pericardial coelomic cavity
Pharyngeal pouch 1
Pharyngeal pouch 2
Pharyngeal pouch 3
Post. amniotic fold
Post. cardinal V.
Post. end of amnion
Post. end of otic vesicle
Post. intestinal portal
Post. tela choroidea
Post. tuberculum
Proamnion
Rathke's pouch
R. omphalomesenteric A.
R. omphalomesenteric V.
R. vitelline V.
Roof of pharynx
Sensory layer
Single dors. aorta
Sinus venosus
Somatopleura
Somite
Spinal cord
Spinal gang.
Splanchnopleura
Stalk of somite
Stomodeum
Tapetum nigrum
Telencephalon
Telocoele
Torsion
Trabeculae carneae
Transverse septum
Truncus arteriosus
Umbilical V.
Velum transversum
Vent. aorta
Vent. liver diverticulum
Vent. mesentery
Ventricle
Vitelline A.
Vitelline V.
Yolk sac

DEVELOPMENT OF
THE CHICK EMBRYO VI

72-HOUR CHICK EMBRYO
Acoustic gang. N. 8
Allantoic A.
Allantoic cavity
Allantoic vessels
Allantois
Amnion
Amniotic cavity
Amniotic fold

nt. cardinal V.
nt. cardinal V. cephalic portion
nt. cardinal V. thoracic portion
nt. Cerebral A.
nt. end of optic cup
nt. intestinal portal
nt. vitelline V.
ortic arch 1
ortic arch 2
ortic arch 3
ortic arch 4
pex cordis
pical ridge
trium
asilary A.
ranch of ant. cardinal V.
ranchial arch 1
ranchial arch 2
ranchial arch 3
ranchial arch 4
ranchial groove 1
ranchial groove 2
ronchus
auda
audal aorta
entral canal
ephalic flexure
erebellar V.
horio-amniotic raphe
horion
horoid fissure
loaca
loacal membrane
osing plate
omm. cardinal V.
orpus vitreum
ielocoele
iencephalon
ividing dors. aorta
ors. aorta
ors. liver diverticulum
ors. mesentery
ors. spinalis V.
uctus venosus
uodenum
ctoderm
mbryonic coelom
ndocardium
nd of subcaudal pocket
ndolymphatic duct
pimyocardium
piphysis
sophagus
xt. carotid A.
xt. naris
xtra embryonic coelom
acial N. 7
cial V.
eniculate gang N. 7
omerulus

Glossopharyngeal N. 9
Hindgut
Hind limb bud
Hepatic A.
Hepatic cord
Hepatic V.
Hyoid arch
Infundibulum
Int. carotid A.
Intersegmental A.
Intersegmental V.
Isthmus
Jugular V.
Lamina terminalis
Lat. basilary A.
Lat. body fold
Lens
L. omphalomesenteric V.
L. post. vitelline V.
L. wing bud
Limb bud
Liver
Liver cord
Lung bud
Mandibular arch
Mandibular branch of trigeminal
  N. semilunar gang. N. 5
Marginal layer
Maxillary branch of trigeminal
  N. semilunar gang. N. 5
Maxillary process
Mesencephalon
Mesenchyme
Mesocardium
Mesocoele
Mesoderm
Mesonephric duct
Mesonephric tubule
Metencephalon
Metocoele
Mouth
Myelencephalon
Myelocoele
Myocoele
Myotome
Neuromere
Neuromere 1 of myelencephalon
Neuromere 2
Neuromere 5
Notochord
Oculomotor N. 3
Olfactory pit
Olfactory placode
Omphalomesenteric A.
Omphalomesenteric V.
Ophthalmic branch of trigeminal
  N. semilunar gang. N. 5
Optic cup
Optic stalk
Orafice of mesonephric duct

Otic vesicle
Paired dors. aorta
Pericardial cavity
Pericardium
Petrosal gang. N. 9
Pharyngeal pouch 1
Pharyngeal pouch 2
Pharyngeal pouch 3
Pharyngeal pouch 4
Pharynx
Postanal gut
Post. cardinal V.
Post. cerebral A.
Post. end of atrium
Post. end of optic cup
Post. end of otic vesicle
Post. intestinal portal
Post. tela choroidea
Post. vitelline V.
Proamnion
Proctodeum
Pronephric duct
Pulmonary A.
Pulmonary cavity
Pulmonary V.
Rathke's pouch
Recessus opticus
Renal A.
R. omphalomesenteric A.
R. omphalomesenteric V.
R. post. vitelline V.
R. wing bud
Roof of allantois
Roof of pharynx
Sclerotome
Semilunar gang. N. 5
Sensory layer
Single dors. aorta
Sinus venosus
Somite
Spinal cord
Spinal gang.
Subcardinal V.
Subcaudal pocket
Subintestinal V.
Tapetum nigrum
Telencephalon
Telocoele
Thyreoglossal duct
Torsion
Trabeculae carneae
Transverse septum
Trans. spinalis A.
Trans. spinalis V.
Truncus arteriosus
Umbilical V.
Urogenital sinus
Velum transversum
Vent. liver diverticulum
Vent. mesentery

333

Vent. nerve root
Ventricle
Vertebral A.
Vitelline A.
Vitelline V.
Wing bud
Yolk sac

DEVELOPMENT OF
THE CHICK EMBRYO VII

96-HOUR CHICK EMBRYO
Abducens N. 6
Acoustic gang. N. 8
Acoustic gang. N. 8, vestibulocochlearis
Allantoic A.
Allantoic stalk
Allantoic V.
Allantoic vessels
Allantois
Amniotic cavity
Amnion
Ant. cardinal V.
Ant. cerebellar. V.
Ant. cerebral A.
Ant. cerebral V.
Ant. vitelline V.
Aortic arch
Apical ridge
Atrium
Axilla
Basilary A.
Branchial A.
Branchial arch 3
Branchial arch 4
Branchial groove 2
Bronchus
Cauda
Caudal aorta
Central canal
Chorion
Choroid fissure
Cloaca
Coelomic cavity
Comm. cardinal V.
Corpus vitreum
Diencephalon
Dors. aorta
Dors. mesentery
Ductus venosus
Duodenum
Ectoderm
Endolymphatic duct
Ependymal layer
Epiglottis
Epiphysis
Esophagus
Ext. carotid A.
Ext. naris

Eye
Facial V.
Geniculate gang. facial N. 7
Glomerulus
Hepatic A.
Hepatic V.
Hyoid (branchial) arch 2
Infundibulum
Iliac A.
Int. carotid A.
Intersegmental A.
Intersegmental V.
Intestine
Isthumus
Jugular gang. N. 10
Jugular gang. N. 10 Vagus
Jugular V.
L. omphalomesenteric A.
L. omphalomesenteric V.
L. post. vitelline A.
L. post. vitelline V.
L. vitelline A.
Lens
Limb bud
Liver
Lung
Lung bud
Mandibular arch
Mandibular (branchial) arch 1
Mandibular N. 5
Mantle layer
Marginal layer
Marginal V.
Maxillary N. 5
Maxillary process
Mesencephalon
Mesocoele
Mesoderm
Mesonephric duct
Mesonephric tubule
Mesonephros
Metencephalon
Mouth cavity
Myelencephalon
Myotome
Neural tube
Neuromere
Notochord
Oculomotor N. 3
Olfactory N. 1
Omphalomesenteric A.
Omphalomesenteric V.
Ophthalmic N. 5
Optic muscle
Optic recess
Optic stalk
Otic vesicle
Pancreas
Pericardium
Petrosal gang. N. 9

Petrosal gang. N. 9, glossopharyngeal
Pharyngeal pouch
Pharyngeal pouch 1
Pharynx
Post. cardinal V.
Post. cerebellar V.
Post. cerebral A.
Post. cerebral V.
Post. tela choroidea
Post. vitelline V.
Proctodeum
Pulmonary A.
Pulmonary cavity
Rathke's pouch
R. omphalomesenteric V.
R. post. vitelline V.
R. vitelline A.
Semilunar gang. N. 5
Semilunar gang. N. 5, trigeminal N.
Sensory layer
Sinus venosus
Somite
Spinal accessory N. 11
Spinal cord
Spinal gang.
Spinal N.
Spinalis dors. V.
Spinalis vent. A.
Stomach
Subclavian A.
Subclavian V.
Sympathetic gang.
Tail
Tapetum nigrum
Telencephalon
Thyreoglossal duct
Trabeculae carneae
Trachea
Trans. spinalis A.
Truncus arteriosus
Umbilical V.
Urogenital sinus
Vagus N. 10
Velum tranvsersum
Vent. nerve root
Ventricle
Vertebral A.
Wing bud
Yolk sac

DEVELOPMENT OF
THE CHICK EMBRYO VIII

HIGH MAGNIFICATION, 72- AND
96-HOUR CHICK EMBRYO
Alar plate
Ant. eye chamber
Basal plate

Basilary A.
Central canal
Choroid fissure
Dermatome
Dors. arota
Ectoderm
Ependymal layer
Epithelial layer
Eye
Lens
Mantle layer
Marginal layer
Mesencephalon
Mesoderm
Mesonephric duct
Mesonephric tubule
Mesonephros
Mitotic figure
Myelencephalon
Myotome
Notochord
Post. cardinal V.
Proliferative zone
Sclerotome
Sensory layer
Somite
Spinal cord
Spinal gang.
Subcardinal V.
Trans. spinalis A.
Tapetume nigrum

ATLAS OF DEVELOPMENTAL ANATOMY

PART III

MAMMALIAN DEVELOPMENT I
RAT TESTIS
Arteriole
Artery
Basal lamina
Basement membrane
Blood vessel
Dividing spermatocytes
Dividing spermatogonia
Early spermatid
Fibroblast
Interstitial cell
Lumen of seminiferous tubule
Maturing spermatid
Nucleolus
Nucleus of Sertoli cell
Primary spermatocyte
Primary spermatocyte, anaphase
Primary spermatocyte, diakinesis
Primary spermatocyte, diplotene
Primary spermatocyte, metaphase
Primary spermatocyte, mitotic figure

Primary spermatocyte, pachytene
Primary spermatocyte, prometaphase
Primary spermatocyte, prophase,
   leptotene
Primary spermatocytes, prophase,
   zygotene
Primary spermatocyte, telophase
Residual body of Regand
Secondary spermatocyte
Seconsary spermatocyte, second
   metaphase
Sertoli cell
Sperm
Spermatid
Spermatogonia
Spermatozoa
Tunica albuginea
Vein
Venule

CAT OVARY
Antrum
Artery
Atretic corpus luteum
Atretic follicle
Blood vessel
Call-Exner vacuole
Cortex
Cumulus oophorus
Follicle cut under the oocyte
Graafian follicle
Medulla
Mesovarium
Oocyte
Ovarian epithelium
Ovarian stroma
Peritonial mesothelium
Primary follicle
Primordial follicle
Secondary follicle
Tertiary follicle
Theca interna
Theca externa
Tunical albuginea
Vein
Zona pelucida

RAT OVARY, CORPUS LUTEUM
Adipose tissue
Artery
Blood vessel
Corpus luteum
Cortex
Graafien follicle
Medulla
Ovarian epithelium
Peritoneum
Primary follicle
Tunica albuginea

Vein

HUMAN OVARY, CORPUS LUTEUM. CORPUS ALBICANS
Artery
Blood clot
Corpus albicans
Corpus luteum
Medulla
Tunica albuginea
Vein

PREGNANT UTERUS OF THE RAT
Amnion
Ant. cardinal V.
Aortic arch
Caudal aorta
Caudal V.
Central A.
Chorionic plate
Cornea
Decidua
Dors. aorta
Dors. nerve root
Ext. carotid A.
Exocoele
First pharyngeal pouch
Giant cell
Glossopharyngeal N.
Intersegmental V.
Labyrinth
Lamina terminalis
Lat. ventricle
Lens
Maternal blood
Membrana granulosa
Mcsometrium
Notochord
Optic stalk
Oral cavity
Pharynx
Placenta
Second pharyngeal pouch
Sensory layer
Smooth chorion
Somite
Spinal cord
Spinal gang.
Tapetum nigrum
Telencephalon
Umbilical A.
Umbilical V.
Uterin epithelium
Uterin longitudinal and circular muscle
Vagus N.
Vent. nerve root
Vertebral A.

MAMMALIAN DEVELOPMENT II
10 MM. PIG EMBRYO

Abducens 6 N.
Acoustic gang. 8 N.
Acoustic and geniculate ganglia
   8 and 7 Nn.
Alar plate
Allantoic sac
Allantoic stalk
Allantois
Amnion, somatopleura (mesoderm and
   ectoderm)
Amniotic cavity
Anastomosis of basilary A. and
   int. carotid A.
Anastomosis, basilary A. and
   post. cerebral A.
Anastomosis, L. and R. umbilical Vv.
Anastomosis, post. cardinal and
   supracardinal Vv.
Anastomosis, post. vena cava and
   post. cardinal Vv.
Anastomosis, post. vena cava and
   subcardinal Vv.
Anastomosis, post. vena cava and
   supracardinal Vv.
Anastomosis, sup. and inf. mesenteric Aa.
Ant. cardinal V.
Ant. cardinal V. branches
Ant. cardinal V., cephalic portion
Ant. cardinal V., thoracic portion
Ant. cerebral A.
Ant. edge of r. ventricle
Ant. edge of umbilicus
Ant. limb bud
Ant. termination of vent. aorta
Aortic arch 2
Aortic arch 3
Aortic arch 4
Aortic arch 6
Aortic arch 6, pulmonary A.
Aorticopulmonary septum
Apical ridge
Arytenoid process
Ascending aorta
Atrio-ventricular canal
Atrium
Axillary N.
Basal plate
Basilary A.
Bifurcation of trachea
Body wall
Brachial N.
Brachial plexus
Branch, ant. cardinal V.
Branch, basilary A.
Branchial arch 3
Branchial arch 4
Branchial groove 2

Branchial groove 3
Branchial groove 4
Bronchus
Cardiac portion of stomach
Cauda
Caudal aorta
Caval plica
Central canal
Cerebellar Aa.
Cervical flexure
Cervical sinus, branchial groove 3
Choroid fissure
Cloaca
Cloacal membrane
Closing plate
Closing plate 1
Closing plate 2
Closing plate 3
Coeliac axis
Coelomic cavity
Colic A.
Colic V.
Colon
Comm. cardinal V.
Comm. trunk of aorta and
   truncus pulmonaris
Copula
Cornea
Corneal ectoderm
Cystic duct
Dermatome
Diencephalon
Diocoele
Divided post. vena cava V.
Dors. aorta
Dors. aorta fused
Dors. lobe of pancreas
Dors. mesentery
Dors. mesogastrum
Dors. nerve root
Dors. pancreatic duct
Dors. (inter-) segmental A.
Dors. (inter-) segmental V.
Ductus choledochus
Ductus venosus
Duodenum
Ectoderm
Endocardial cushion
Endolymphatic duct
Ependymal layer
Epiglottis
Epiploic
Esophagus
Ext. carotid A.
Ext. jugular V.
Eye
Eye muscle primordium
Facial 7 N.
Facial V.

Falciforme ligament
Fibrous capsule
First branchial groove
First neuromere of myelencephalon
Forelimb
Froriep's gang., spinal accessory 11 N.
Gall bladder
Geniculate gang., facial 7 N.
Genital ridge
Glomerulus
Glossopharyngeal 9N.
Greater omentum
Head ectoderm
Head mesoderm
Hepatic cords
Hepatic duct
Hepatic portal V.
Hepatic sinus
Hepatic sinusoid
Hindlimb
Hyoid (Branchial) arch 2
Hyomandibular (branchial) groove 1
Hyomandibular groove
Hypoglossal 12 N.
Iliac A.
Iliac V.
Inf. mesenteric A.
Inf. mesenteric (colic) V.
Infundibulum
Interatrial foramen II.
Int. carotid A.
Interventricular foramen
Interventricular septum
Intestinal A.
Intestinal V.
Isthmus
Joined branch of vitelline V.
Jugular gang. 10 N.
Jugular gang. vagus 10 N.
Lamina terminalis
Large intestine
Large intestine, colon
Laryngo-trachial groove
Larynx
Lat. body wall
Lat. hemisphere
Lat. nasal process
L. atrium
L. dors. lobe of liver
L. hindlimb
L. horn of sinus venosus
L. limb bud
L. mesonephric duct
L. subclavian A.
L. umbilical A.
L. umbilical V.
L. umbilical V., abdominal portion
L. umbilical V., umbilical portion
L. uretric bud
L. vent. lobe of liver

L. ventricle
Lens
Lesser omentum
Liver
Lumbar plexus
Lung bud
Mandibular arch
Mandibular (branchial) arch 1
Mandibular N.
Mandibular N. of trigeminal 5 N.
Mammary ridge
Mantle layer
Marginal layer
Maxillary N.
Maxillary process
Median nasal process
Mesencephalon
Mesocoele
Mesocolon
Mesoderm
Mesogastrum
Mesonephric duct
Mesonephric tubule
Mesonephros
Mesorectum
Metanephros
Metencephalon
Metocoele
Mitral valve primordium
Myelencephalon
Myelocoele
Myotome
Nasal placode
Naso-lacrymal groove
Neuromere
Neuromere 1 of myelencephalon
Nodose gang., vagus 10 N.
Notochord
Oculomotor 3 N.
Olfactory pit
Omentum bursa
Ophthalmic N.
Optic cup, eye
Optic cup, retina
Optic stalk
Orafice of Rathke's pouch
Oral cavity
Otic vesicle
Paried dros. aorta
Pectinate muscles
Pericardial cavity
Pericardium
Petrosal gang., glossopharyngeal 9 N.
Petrosal gang. 9 N.
Pharyngeal pouch 1
Pharyngeal pouch 2
Pharyngeal pouch 3
Pharynx
Pigment layer
Pigment layer, tapetum nigrum

Pleuroperitonial fold
Pontine flexure
Postanal gut
Post. cardinal V.
Post. cerebral A.
Post. tela choroidea
Post. vena cava V.
Post. wall of infundibulum
Primordial pelvis
Pulmonary A.
Pulmonary cavity
Pulmonary V.
Pyloric portion of stomach
Ramus communicans
Rathke's pouch
Rectum
Renal A.
Renal V.
Retiforme central A.
Retina
R. atrium
R. auricle
R. dors. lobe of liver
R. hindlimb
R. horn of sinus venosus
R. mesonephric duct
R. metanephric tissue
R. subclavian A.
R. subclavian V.
R. umbilical A.
R. umbilical V.
R. ureter
R. ventricle
Roof of myelencephalon
Roots of facial 7 N. and acoustic 8 N.
Root of 5 N.
Root of oculomotor 3 N.
Root of semilunar gang., trigeminal 5 N.
Sclerotome
Second aortic arch
Second branchial (hyoid) arch
Second branchial groove
Segmental (inter-) V.
Semilunar gang. 5 N.
Sensory layer
Septum primum
Sino-atricular valve
Small intestine
Somatopleura
Somites
Spinal accessory 11 N.
Spinal cord
Spinal gang.
Spinal N.
Spinalis dors. V.
Spinalis vent. A.
Spiral septum
Stalk of allantois
Stalk of allantois enters into
     urogenital sinus

Stalk of yolk sac
Stomach
Subcardinal V.
Subcardinal and supracardinal Vv.,
     anastomosis
Subclavian V.
Succulus
Sulcus limitans
Sup. gang., glossopharyngeal 9 N.
Sup. mesenteric A.
Sup. mesenteric V.
Supracardinal V.
Sympathetic gang.
Tail
Tail somite
Telencephalon
Telocoele
Terminal loop of umbilical hernia
Third aortic arch
Third branchial arch
Third branchial groove
Thyreoid gl.
Trabeculae carneae
Trachea
Trans. septum
Trans. sinus venosus
Trans. spinalis A.
Tricaspid valve primordium
Trochlear 4 N.
Tr. arteriosus
Tr. pulmonaris
Tuberculum impare
Umbilical A.
Umbilical cord
Umbilical hernia
Umbilicus
Urogenital sinus
Utriculus
Utriculus of otic vesicle
Vagus 10 N.
Vent. aorta
Vent. end of optic cup
Vent. end of otic vesicle
Vent. lobe of liver
Vent. lobe of pancreas
Vent. mesentery
Vent. nerve root
Vent. segmental V.
Vent. segmental Vv. from umbilical V.
Ventricle
Vertebral A.
Vertebral A. from dors. aorta
Vitelline A.
Vitelline V.
Wall of atrium
Wall of bronchus
Wall of pharynx
Yolk sac

MAMMALIAN DEVELOPMENT III
MORE-DEVELOPED
10 MM. PIG EMBRYO

Acoustic 8 N.
Alar plate
Ant. cardinal V.
Aortic arch 6, pulmonary A.
Aorticopulmonary septum
Ascending aorta
Basilary A.
Branch, ant. cardinal V.
Branch of basilary A.
Central canal
Coelomic cavity
Comm. cardinal V.
Dors. aorta
Dors. nerve root
Dors. ramus
Dors. (inter-) segmental A.
Dors. (inter-) segmental V.
Ectoderm
Endocardial cushion
Eparterial bronchus
Ependymal layer
Esophagus
Facial 7 N.
Head ectoderm
Head mesoderm
Hypoglossal 12 N.
Interventricular foramen
Interventricular septum
Isthmus
L. atrium
L. auricle
L. horn of sinus venosus
L. ventricle
Mantle layer
Marginal layer
Mesencephalon
Mesocoele
Mesoderm
Mesodermal primordium of otic capsule
Mesonephric tubule
Metencephalon
Metocoele
Myelencephalon
Myelocoele
Nerve root of trochlear 4 N.
Notochord
Oculomotor 3 N.
Pericardial cavity
Pericardium
Petrosal gang. glossopharyngeal 9 N.
Post. cardinal V.
Post. cerebral A.
Post. tela choroidea
Pulmonary A.
Pulmonary cavity
Pulmonary V.
Ramus communicans

R. atrium
R. auricle
R. horn of sinus venosus
R. ventricle
Root of trigeminal 5 N.
Septum primum
Sino-atricular valve
Spinal accessory 11 N.
Spinal gang.
Spinal N.
Spinalis dors. V.
Spinalis vent A.
Succulus
Sulcus limitans
Sympathetic gang.
Thoracic N.
Trabeculae carneae
Trachea
Trans. spinalis A.
Tricaspid valve primordium
Tr. arteriosus
Tr. pulmonaris
Vagus 10 N.
Vent. nerve root
Vent. ramus
Vent. segmental V.